Readings in Mathematics

BOOK 2

EDITED BY IRVING ADLER

GINN AND COMPANY
A XEROX COMPANY

Acknowledgements

Grateful acknowledgement is made to the following authors and publishers for permission to use copyrighted materials:

To Nathan Ainsworth for his article "An Introduction to Sequence: Elementary School Mathematics and Science Enrichment." Reprinted from the *Arithmetic Teacher*, February, 1970 (vol. 17, pp. 143-45), © 1970 by the National Council of Teachers of Mathematics. Used by Permission.

To Aldus Books Limited, London for "Square and Circle" and "The Pendulum and Galileo" by Lancelot Hogben from *Wonderful World of Mathematics*. Copyright © 1968 by Aldus Books Limited. Copyright © 1955 by Rathbone Books Limited, London. Reprinted by permission.

To Chatto and Windus Ltd, London for "On Being the Right Size" from *Possible Worlds* by J. P. S. Haldane. Copyright, 1928 by Harper & Row, Publishers, Inc.; renewed, 1956 by J. B. S Haldane.

To The Christian Science Publishing Society for the article "Draft Unfair to December?" by Eric Burgess, which appeared January 5, 1970. Also for "UN examines population boom" which appeared December 1, 1969. Reprinted by permission from The Christian Science Monitor © 1970, The Christian Science Publishing Society. All rights reserved.

To the Clarendon Press for the extract from *A History of Greek Mathematics* by Sir Thomas L. Heath, 1921. Reprinted by permission of The Clarendon Press, Oxford.

To Doubleday and Company, Inc. for "The Birthday Problem," From LADY LUCK by Warren Weaver, copyright © 1963 by Educational Services, Incorporated. Reprinted by permission of Doubleday & Company, Inc.

To M. C. Escher for permission to reproduce the prints 35, 68, and 69 from the book the *Graphic Work of M. C. Escher* and the prints 6 and 14 including the "Preface" from the book *Symmetry Aspects of M. C. Escher's Periodic Drawings* by C. H. MacGillvary.

To Mrs. Helen Spurway Haldane for "On Being the Right Size" from *Possible Worlds* by J. P. S. Haldane.

To Harper & Row, Publisher, Incorporated for the selection from *Life on the Mississippi* by Mark Twain. Also for "On Being the Right Size" from *Possible Worlds* by J. P. S. Haldane. Copyright, 1928 by Harper & Row, Publishers, Inc.; renewed, 1956 by J. B. S. Haldane. Reprinted by permission of the publishers.

To Heinemann Educational Books Ltd, London for "The Birthday Problem." From LADY LUCK by Warren Weaver, copyright ©1963 by Educational Services, Incorporated. Reprinted by permission.

To Thomas J. Hill, Editor of *The Mathematics Student Journal* for "The Hare and the Tortoise — and Other Paradoxes" by James T. Fey, which appeared in the May, 1968 issue of *The Mathematics Student Journal*, a publication of The National Council of Teachers of Mathematics.

To International Union of Crystallography for permission to use the "Preface." Reprinted from *Symmetry Aspects of M. C. Escher's Periodic Drawings* by C. H. MacGillavry published by A. Oosthoek's Uitgivers N.V., for the International Union of Crystallography.

To John Day Company, Inc. for "Building Blocks of the Universe," copyright ©1963 by Irving Adler. Reprinted from Inside the Nucleus by Irving Adler, by permission of the John Day Company, Inc. publisher. Also for "From Decimal Fraction to Common Fraction," "From Repeating Decimal to Common Fraction," "Nonterminating Decimals," "Rates, Paradoxes, and Fallacies," and exercises 14-17 and answers, copyright ©1964. Reprinted from A New Look at Arithmetic by Irving Adler, by permission of the John Day Company, Inc. publisher. Also for "The Game of 'As If'," copyright ©1964 by Irving Adler. Reprinted from Logic for Beginners by Irving Adler, by permission of the John Day Company, Inc. publisher. Also for "Space Filling Figures," copyright ©1966 by Irving Adler. Reprinted from A New Look at Geometry by Irving Adler, by permission of John Day Company, Inc. publisher. Also for "Splitting the Ring," copyright ©1957 by Irving Adler. Reprinted from The Magic House of Numbers by Irving Adler, by permission of the John Day Company, Inc. publisher. Also for "Why Dust Floats," copyright ©1958 by Irving Adler. Reprinted from Dust by Irving Adler, by permission of the John Day Company, Inc. publisher.

To McGraw-Hill Book Company for both "Fermat's Method of Factoring" and "Indeterminate Problems" from *Number Theory and Its History* by Oystein Ore. Copyright 1948 by McGraw-Hill, Inc. Used with permission of McGraw-Hill Book Company.

To Northwestern University Press for both "The Surfaces and Volumes of Similar Figures" and "The Rhythm of the Pendulum" by Galileo Galilei, translated by Henry Crew and Alfonso De Salvio. Reissued in 1939 by THH Editorial Board of Northwestern University Studies. Reprinted by permission of Northwestern University Press, Evanston, Illinois.

To Oxford Press for "Fragile Animals Under High Pressure." From *The Sea Around Us* by Rachel L. Carson. Copyright ©1950, 1951, 1961 by Rachel L. Carson. Reprinted by permission of Oxford University Press, Inc.

To Evelyn Singer for the selections from *The Story of Mathematics — Geometry for the Young* by Hy Ruchlis and Jack Engelhardt. Copyright ©1958 by Z. E. Harvey, Inc. Published by Harvey House.

To The Viking Press, Inc. for "The Law of Disorder." From ONE, TWO, THREE . . . INFINITY by George Gamow. Copyrighted ©1947 by George Gamow. Reprinted by permission of The Viking Press, Inc.

Acknowledgements continued on page 188.

Contents

TO THE READER

This book was put together for your enjoyment. If, as you read for pleasure, you learn something new at the same time, that good too.

The selections in this book show you mathematics in use — in everyday life, in science, and in imaginative literature. The mathematics involved is elementary mathematics. Elementary does not mean trivial or unimportant. Elementary means fundamental and simple. In fact, a major goal of this book is to show that it is often possible to solve interesting, important, and even complicated problems by making skillful use of a few simple ideas.

The selections were taken from many sources. They include articles from newspapers and magazines; excerpts from the books of contemporary authors who are well known for the skill with which they present the ideas of mathematics and science; and quotations from the original writings of great scientists and mathematicians: Euclid, Archimedes, Galileo, Newton, Gauss, and Pascal.

Begin by browsing. Glance through the table of contents. Thumb through the book and look at the pictures. Read some of the short paragraphs that introduce the selections. Skim lightly through anything that catches your attention. After tasting the flavor of the book in this way, you will be ready to go on with a more systematic reading of it. In this next phase of your reading, remember that mathematics is not chiefly a spectator sport. You can enjoy it best as something you *do* rather than as something you *watch*. If a proposition is proved, you appreciate it most if you proved it. If a discovery is made, you enjoy it most if you discovered it. If a problem is solved, it thrills you most if you solved it. For this reason, it is suggested that you read this book as a participant, not as a spectator. Read it with paper and pencil at hand. When you read a selection through for the first time, read it for an overview only. Then read it through a second time for details. Use your pencil and paper to work through some of the details by yourself. If an equation is being solved, and some steps have been omitted, supply the missing steps. If a diagram is being described, draw the diagram yourself, step by step to match the description. If exercises are suggested, work them out.

1

The selections may be read in any order. However, some ...re connected with each other will be enjoyed most if they ...ad in sequence. For this reason they have been grouped to-...er. To help you understand some of the ideas developed, there ...notes in the margin or interpolations in square brackets. Read ...ese even if they slow you down. Remember that in mathematical ...eading speed is not important, but understanding is.

Now, get your pencil and paper ready and enjoy yourself.

Irving Adler

Song of the Major General

from *The Pirates of Penzance*
by *W. S. Gilbert*

GEN. I am the very model of a modern Major-General,
I've information vegetable, animal, and mineral,
I know the kings of England, and I quote the fights historical,
From Marathon to Waterloo, in order categorical;
I'm very well acquainted too with matters mathematical,
I understand equations, both the simple and quadratical,
About binomial theorem I'm teeming with a lot o' news—
With many cheerful facts about the square of the hypotenuse.

ALL. With many cheerful facts, etc.

GEN. I'm very good at integral and differential calculus,
I know the scientific names of beings animalculous;
In short, in matters vegetable, animal, and mineral,
I am the very model of a modern Major-General.

PART 1

The Expected

12345

The Law of Disorder

by George Gamow

If the outcome of an experiment depends on chance, and may be any one of several possible events, we do not know what the outcome of one experiment will be, but we may be able to predict how many of each possible outcome to expect in many experiments. The well-known physicist George Gamow explains what this means in his book, One, Two, Three . . . Infinity, *using as an illustration the code problem in Edgar Allan Poe's story, "The Gold Bug."*

It cannot be repeated too often that if we calculate the probabilities of different events according to the given rules and pick out the most probable of them, we are not at all sure that this is exactly what is going to happen. Unless the number of tests we are making runs into thousands, millions or still better into billions, the predicted results are only "likely" and not at all "certain." This slackening of the laws of probability when dealing with a comparatively small number of tests limits, for example, the usefulness of statistical analysis for deciphering various codes and cryptograms which are limited only to comparatively short notes. Let us examine, for example, the famous case described by Edgar Allan Poe in his well-known story "The Gold Bug." He tells us about a certain Mr. Legrand who, strolling along a deserted beach in South Carolina, picked up a piece of parchment half buried in the wet sand. When subjected to the warmth of the fire burning gaily in Mr. Legrand's beach hut, the parchment revealed some mysterious signs written in ink which was invisible when cold, but which turned red and was quite legible when heated. There was a picture of a skull, suggesting that the document was written by a pirate, the head of a goat, proving beyond any doubt that the pirate was none other than the famous Captain Kidd, and several lines of typographical signs apparently indicating the whereabouts of a hidden treasure (see Figure 87).

We take it on the authority of Edgar Allan Poe that the pirates of the seventeenth century were acquainted with such typographical signs as semicolons and quotation marks, and such others as: ‡, +, and ¶.

Being in need of money, Mr. Legrand used all his mental powers in an attempt to decipher the mysterious cryptogram and

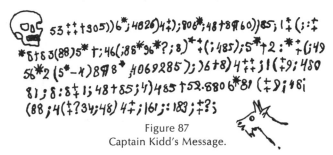

Figure 87
Captain Kidd's Message.

finally did so on the basis of the relative frequency of occurrence of different letters in the English language. His method was based on the fact that if you count the number of different letters of any English text, whether in a Shakespearian sonnet or an Edgar Wallace mystery story, you will find that the letter "e" occurs by far most frequently. After "e" the succession of most frequent letters is as follows:

a, o, i, d, h, n, r, s, t, u, y, c, f, g, l, m, w, b, k, p, q, x, z ✱

✱ **For the average frequency with which each letter occurs, see the table on page 8.**

By counting the different symbols appearing in Captain Kidd's cryptogram, Mr. Legrand found that the symbol that occurred most frequently in the message was the figure 8. "Aha," he said, "that means that 8 most probably stands for the letter e."

Well, he was right in this case, but of course it was only *very probable* and not at all certain. In fact if the secret message had been "You will find a lot of gold and coins in an iron box in woods two thousand yards south from an old hut on Bird Island's north tip" it would not have contained a single "e"! But the laws of chance were favorable to Mr. Legrand, and his guess was really correct.

Having met with success in the first step, Mr. Legrand became overconfident and proceeded in the same way by picking up the letters in the order of the probability of their occurrence. In the following table we give the symbols appearing in Captain Kidd's message in the order of their relative frequency of use:

I		II	III
Of the character 8 there are	33	e ⟷	e
;	26	a	t
4	19	o	h
‡	16	i	o
(	16	d	r
*	13	h	n
5	12	n	a
6	11	r	i
†	8	s	d
1	8	t	
0	6	u	
g	5	y	
2	5	c	
i	4		
3	4	g ⟵	g
?	3	l	u
¶	2	m	
—	1	w	
·	1	b	

The first column [II] on the right contains the letters of the alphabet arranged in the order of their relative frequency in the English language. Therefore it was logical to assume that the signs listed in the broad column [I] to the left stood for the letters listed opposite them in the first narrow column [II] to the right. But using this arrangement we find that the beginning of Captain Kidd's message reads: *ngiisgunddrhaoecr . . .*

No sense at all!

What happened? Was the old pirate so tricky as to use special words that do not contain letters that follow the same rules of frequency as those in the words normally used in the English language? Not at all; it is simply that the text of the message is not long enough for good statistical sampling and the most probable distribution of letters does not occur. Had Captain Kidd hidden his treasure in such an elaborate way that the instructions for its recovery occupied a couple of pages, or, still better an entire volume, Mr. Legrand would have had a much better chance to solve the riddle by applying the rules of frequency.

If you drop a coin 100 times you may be pretty sure that it will fall with the head up about 50 times, but in only 4 drops you may have heads three times and tails once or vice versa. To make a rule of it, *the larger the number of trials, the more accurately the laws of probability operate.*

Since the simple method of statistical analysis failed because of an insufficient number of letters in the cryptogram, Mr. Legrand had to use an analysis based on the detailed structure of different words in the English language. First of all he strengthened his hypothesis that the most frequent sign *8* stood for *e* by noticing that the combination 88 occurred very often (5 times) in this comparatively short message, for, as everyone knows, the letter e is very often doubled in English words (as in: *meet, fleet, speed, seen, been, agree, etc.*). Furthermore if *8* really stood for *e* one would expect it to occur very often as a part of the word "the." Inspecting the text of the cryptogram we find that the combination; *48* occurs seven times in a few short lines. But if this is true, we must conclude that *;* stands for *t* and *4* for *h*.

We refer the reader to the original Poe story for the details concerning the further steps in the deciphering of Captain Kidd's message, the complete text of which was finally found to be: "A good glass in the bishop's hostel in the devil's seat. Forty-one degrees and thirteen minutes northeast by north. Main branch seventh limb east side. Shoot from the left eye of the death's head. A bee-line from the tree through the shot fifty feet out."

The correct meaning of the different characters as finally deciphered by Mr. Legrand is shown in the second column [III] of the table on page 8, and you see that they do not correspond exactly to the distribution that might reasonably be expected on the basis of the laws of probability. It is, of course, because the text is too short and therefore does not furnish an ample opportunity for the laws of probability to operate. But even in this small "statistical sample" we can notice the tendency for the letters to arrange themselves in the order required by the theory of probability, a tendency that would become almost an unbreakable rule if the number of letters in the message were much larger.

Hamming It Up— Mathematically

by Dale M. Shafer

Indiana University of Pennsylvania

Is Morse code the best possible telegraph code using dots and dashes? The following selection, adapted from an article that appeared in the February 1970 issue of School Science and Mathematics, *tells you how to answer this question.*

Morse code is designed for the rapid transmission of messages by radiotelegraphy. The letters are formed by a combination of no more than four dots and/or dashes as indicated in column (2) of the table on page 18. To form a dash the telegraph key is depressed for a time unit three times as long as for a dot. The element space between dots and dashes in the same letter has the same time unit length as a dot. The relationship between dots, dashes and element spaces is indicated in the diagram below for the letter "r" whose Morse character is "• — •".

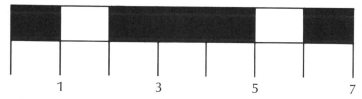

Thus, r has a time unit length of seven. The time unit length for all twenty-six letters is entered in column (3) of the table.

[The cost of sending a message depends on the number of time units in the length of the message. This in turn depends on how often each letter of the alphabet occurs in the message. The number of times each letter of the alphabet occurs, on the average, in an English message of one hundred letters is shown in the table on the next page.

10

	Frequency of appearance in average message of one hundred letters
Letters	Number of Appearances
E	13
T	10
AONRIS	7
H	5
DLFCMU	3
GYPWB	2
VKXJQZ	½

The frequencies shown in column 2 of this table are repeated in column (4) of the table on page 18. If, in the latter table, we multiply each number in column (4) by the number next to it in column (3), we get the number of time units required for each letter of the alphabet in the average message of one hundred letters. These products are recorded in column (5). The total of the numbers in column (5) is the time unit length of the average message of one hundred letters.

From the point of view of economy of use, the best possible code is one that gives the smallest possible value for the time unit length of the average message of one hundred letters. Is the Morse code the best possible code from this point of view? To answer this question, see if you can change the code in a way that reduces the time unit length required for the average message of one hundred letters.]

The most important game of chance in the United States is the draft lottery. The lottery is supposed to pick at random and thus arrange in random order the different days of the year. Then the order in which a draftee's birthday is picked will be the order in which he is called for military service. The draft lottery that took place in December 1969 was criticized as being unfair because it did not really pick the birth dates in a random manner. To meet this criticism, a new system for picking the birth dates was introduced with the lottery that took place in June 1970. The next two selections are newspaper stories. The first one reports the criticism that was made of the old lottery system. The second one describes the new lottery system and its results.

Draft Unfair to December?

by Eric Burgess

Staff correspondent of The Christian Science Monitor

STATISTICIAN CRITICIZES U.S. LOTTERY SYSTEM, CHARGES IT FAILS TO BE FULLY RANDOM AND SELECTIVE

When is a lottery not a lottery?

When it is not a random selection.

A lottery is generally defined as a plan for the distribution of prizes by chance. And Webster defines chance as "something that happens unpredictably without any discernible human intention or direction and in dissociation from any observable pattern, casual relation, natural necessity, or providential dispensation."

The draft lottery to select which male teen-agers should have the "prize" of serving in the armed forces was intended as a matter of chance — a random selection in which each number selected has the same probability of being selected as any other number in the set.

The numbers were related to birthdays to select the draftees.

From the *Christian Science Monitor,* January 5, 1970.

DISAGREEMENT VOICED

But Prof. Fred T. Haddock of the University of Michigan, Ann Arbor, says the lottery in December was far from being random.

Professor Haddock is a radio astronomer who spends most of his time statistically analyzing numbers that represent data on distant galaxies and processes occurring in deep space.

"We have to make decisions continually on our measurements," he said in a telephone interview. "And we always apply statistical tests on everything we do."

A casual look at the draft-lottery numbers shows that something is wrong, he said. And a full analysis shows a serious discrepancy.

"Other people have made similar calculations, but they have been reluctant to come out and say that they have doubts about the lottery," he adds.

Professor Haddock says there is an increasing number of men liable to be drafted as their birth dates fall later in the year. This trend can be seen by plotting the average monthly draft number from January through December. The plot gives a clear trend of increasing draft risk with later dates of birth. [See page 14]

LATE-MONTH BIAS SEEN

"The odds against this happening in a true random selection is more than 70,000 to 1," he said.

"The closest set of odds I can find to this 70,000 to 1 is the chance of taking a freshly shuffled deck of cards and dealing off five cards from the top in a straight flush, or the chance of flipping a penny and have it come up heads 16 times in a row."

In other words the lottery is strongly biased. It makes young men with birthdays toward the end of the year more likely to be drafted.

"It is a big blunder, and it can't be swept under the rug as some people in Washington appear to be trying to do."

"It makes the lottery look as though it is just a pacifier," he added.

DRAFT ELIGIBLES GO TO COURT

Already there have been reactions to Professor Haddock's calculations. "The National Academy of Sciences has been asked to analyze the methods used in the draft lottery," he said.

In the United States District Court in Madison, Wis., 13 registrants petitioned the court to have a restraining order issued on their drafting because of the unfairness of the draft, he added.

An analysis of the draft results leads Professor Haddock to conclude that the trend is just what would be expected if the birth dates were dropped into the bowl month by month and then were selected without mixing.

"It appears that the capsules were just dropped into the selection bowl, and the people doing this assumed that because the capsules are round they would be randomly distributed."

The draft lottery can be righted easily, said Profesor Haddock. "Men born in November and December with draft numbers below 184 should be given a new deal. This could be done by having their 47 birthdays redrawn from a new lottery. Then their new order numbers should be mutiplied by a factor to give them an even spread from 1 to 366. This would not unduly affect the chances of the other 87 percent of the young men.

"Without this or a similar remedy, the men born in November and December will be subjected to an unfavorably biased treatment in opposition to the intent and spirit of the lottery," he concluded.

PROPER MIX SOUGHT

The selective-service people could easily have made this a proper lottery, the professor said. A call to anyone in statistics — such as at the Department of Agriculture where there are hundreds of experts — could have immediately produced recommendations on how best to mix the capsules for randomness.

December 1969 Lottery Monthly Average Draft Numbers	
Month	Average Draft Number
January	201
February	203
March	226
April	204
May	208
June	192
July	188
August	173
September	157
October	182
November	149
December	121

July 9 Is 'No. 1'

United Press International

This article reports the first use of the revised system: The Draft Lottery of June 1970.

WASHINGTON (UPI) — Nineteen-year-olds born on July 9, 1951, were assigned the No. 1 callup for induction next year in the national draft lottery held today.

The birthdate was the 11th picked in the fateful drawing to decide the military future of 600,000 19-year-olds, and a capsule drawn from a separate drum gave July 9 the No. 1 position in the priority order of Selective Service callup.

Thus, under the two-capsule system, Sept. 16 was the first birthdate drawn, but youths born on that date were assigned an induction priority of 139.

The 10 birthdates drawn prior to July 9 ranged in priority order from 5 for Oct. 21 to 257 for July 12 through the luck of the draw.

After a momentary mechanical hitch which delayed the drawing by 15 minutes, the second national lottery began smoothly under a scientifically devised system aimed at making the order of selection as truly random as possible.

Two capsules were drawn from two separate drums — one bearing a date of birth; the other determining the priority a youth born on that day would be called for induction.

The randomness of the selection was indicated by the fact that a capsule containing a date for each of the 12 months of the year had been pulled on the 34th draw.

The lottery started about 15 minutes late when a hand crank used to rotate the drum containing the induction priority

(from *The Bennington Banner*, July 1, 1970)

capsules broke off, necessitating immediate repairs. The maker of the drums, C. D. Nelson of nearby Arlington, Va., and his wife, worked hastily with government mechanics to put the drum in operating order again.

The capsules were plucked from the two drums by youthful members of the Selective Service's Youth Advisory Committee, many of them subject to the draft through last year's drawing.

For the second time in seven months the Selective Service System turned to a random drawing of birthdates to determine which youths turning 19 this year would be called for possible induction. The new lottery affected only the estimated 600,000 men born in 1951.

A new presiding officer and a complicated new system were used to assure that the 365 birthdates and the order-of-callup numbers that go with them were as close as possible to true random selection.

Curtis Tarr, a former Pentagon official, replaced the veteran Gen. Lewis B. Hershey as Selective Service director last spring. Under his guidance the lottery procedure was revised to answer scientific complaints that the Dec. 1, 1969 drawing was statistically distorted.

Trying to leave everything to chance, the Selective Service even loaded the birthdate and number capsules into the drums in random order.

THE FIRST 100 DRAWS

WASHINGTON (UPI) — Following are results of the 1970 draft lottery, affecting men becoming 19 this year. Given is the order of the draw, the birthdate and the sequence in which men born on that date in 1951 will be summoned for service:

Order of Draw	Birth Date	Draft Number	Order of Draw	Birth Date	Draft Number
1st	Sept. 16	139	10th	Aug. 18	109
2nd	April 27	235	11th	July 9	1
3rd	Jan. 18	185	12th	Aug. 31	275
4th	Oct. 21	5	13th	March 26	121
5th	Oct. 3	134	14th	Oct. 7	78
6th	July 12	257	15th	Dec. 12	19
7th	April 4	37	16th	Feb. 21	213
8th	Oct. 10	160	17th	May 25	26
9th	May 4	240	18th	May 28	9

Order of Draw	Birth Date	Draft Number	Order of Draw	Birth Date	Draft Number
19th	Feb. 7	25	58th	March 10	150
20th	July 19	316	59th	Oct. 22	36
21st	March 28	95	60th	Jan. 8	116
22nd	June 2	304	61st	Aug. 11	230
23rd	Jan. 30	112	62nd	Oct. 9	302
24th	March 11	317	63rd	June 17	289
25th	July 6	164	64th	Aug. 23	10
26th	March 27	254	65th	Nov. 2	205
27th	April 7	142	66th	Sept. 22	88
28th	Jan. 7	159	67th	Feb. 19	331
29th	May 9	357	68th	July 23	172
30th	Sept. 10	130	69th	Jan. 12	152
31st	Aug. 12	320	70th	Nov. 15	362
32nd	March 31	38	71st	Jan. 27	173
33rd	Jan. 20	211	72nd	Oct. 7	131
34th	Nov. 19	252	73rd	Jan. 24	177
35th	Jan. 13	330	74th	Jan. 14	71
36th	June 3	135	75th	Jan. 22	132
37th	June 7	169	76th	April 15	182
38th	July 24	360	77th	Aug. 8	49
39th	Feb. 25	325	78th	Nov. 1	43
40th	Oct. 8	45	79th	Jan. 19	188
41st	Nov. 8	119	80th	April 29	111
42nd	April 14	202	81st	May 5	301
43rd	April 30	358	82nd	Nov. 30	67
44th	Nov. 4	39	83rd	March 1	14
45th	Dec. 3	110	84th	Oct. 25	17
46th	May 27	122	85th	Aug. 10	359
47th	Nov. 22	253	86th	Nov. 3	294
48th	May 14	40	87th	Aug. 24	274
49th	Aug. 30	167	88th	April 20	118
50th	April 5	124	89th	Sept. 30	18
51st	Aug. 5	64	90th	Dec. 4	305
52nd	April 26	137	91st	March 17	220
53rd	Jan. 6	285	92nd	Feb. 5	97
54th	Dec. 26	80	93rd	March 16	258
55th	Oct. 4	266	94th	Aug. 29	32
56th	Nov. 24	81	95th	July 14	156
57th	April 8	267	96th	April 11	178

Order of Draw	Birth Date	Draft Number	Order of Draw	Birth Date	Draft Number
97th	July 27	85	99th	June 22	307
98th	April 1	224	100th	Feb. 15	201

Exercise: The table on pages 16-17 lists the first one hundred birth dates that were drawn by the new system, and the draft number assigned to each birth date. You can make a rough check of the fairness of the new system in this way: For each month of the year, make a list of the draft numbers assigned to days in that month by this table. Then calculate the average of these draft numbers. If the new system is really fair, the averages for the different months should be almost equal. (To make a more reliable check, you would need the complete list of 365 draws instead of this table that contains only the first one hundred draws.)

English Letter (1)	Morse Character (2)	Time Units (3)	Frequency in 100 letters (4)	Time unit length in 100 letters (5) = (3) × (4)
A	· —	5	7	35
B	— · · ·	9	2	18
C	— · — ·	11	3	33
D	— · ·	7	3	21
E	·	1	13	13
F	· · — ·	9	3	27
G	— — ·	9	2	18
H	· · · ·	7	5	35
I	· ·	3	7	21
J	· — — —	13	$1/2$	$6^{1}/_2$
K	— · —	9	$1/2$	$4^{1}/_2$
L	· — · ·	9	3	27
M	— —	7	3	21
N	— ·	5	7	35
O	— — —	11	7	77
P	· — — ·	11	2	22
Q	— — · —	13	$1/2$	$6^{1}/_2$
R	· — ·	7	7	49
S	· · ·	5	7	35
T	—	3	10	30
U	· · —	7	3	21
V	· · · —	9	$1/2$	$4^{1}/_2$
W	· — —	9	2	18
X	— · · —	11	$1/2$	$5^{1}/_2$
Y	— · — —	13	2	26
Z	— — · ·	11	$1/2$	$5^{1}/_2$
		Total time unit length of average 100 letter message		615

PART 2

The Unexpected

The Birthday Problem

by Warren Weaver

Probability calculations often lead to surprising predictions. That is, when you figure out what may be expected in a situation that depends on chance, the result may be quite unexpected. A good example is the well-known birthday problem, described and solved in the next selecttion, taken from Lady Luck, *by Warren Weaver.*

Dr. Weaver is a distinguished mathematician who has held many important posts in professional organizations and research institutes. He has served as vice-president of both the Rockefeller Foundation and the Alfred P. Sloan Foundation, and in 1954 he was president of the American Association for the Advancement of Science.

Suppose there are n people in a room. What is the probability that at least two of them share the same birthday — the same day of the same month? In our mathematical model we assume that a person is just as likely to be born on one day as another, and we ignore leap years, so that we always have 365 days in a year.

The event that at least two persons share the same birthday is complementary to the event in which they all have distinct birthdays. We start with one person. Whatever day it may happen to be, he *has* a birthday. The probability that person number two has a different birthday is clearly,

$$\frac{364}{365}$$

since it will be different if he was born on any one of the 364 days remaining after we cross off, so to speak, the birthday of the first person. When we advance to the third person, there are 363 permissible days left, so the probability that the third person's birthday differs from that of the first and the second is

$$\frac{363}{365}$$

These are independent events, so the compound probability that number two differs from number one and that number three differs from both number one and two is

$$\frac{364}{365} \times \frac{363}{365}$$

In just the same way the probability of all distinct birthdays for the first four persons is

$$\frac{364 \times 363 \times 362}{365 \times 365 \times 365} = \frac{365 \times 364 \times 363 \times 362}{365 \times 365 \times 365 \times 365}$$

where, in the last-written form, we have inserted an extra 365 both upstairs and downstairs in the fraction (leaving its value unchanged, of course) just so the number of factors will, for convenience in remembering, be equal to the number of persons to which the expression is applied.

It is now easy to generalize to the formula for the probability of all distinct birthdays for n persons. It obviously is

$$\frac{365 \times 364 \times 363 \times 362 \cdots}{365 \times 365 \times 365 \times 365 \cdots}$$

with n factors in both the numerator and denominator of the fraction.

Therefore the probability of the complementary event, namely that at least two persons share the same birthday, is

$$1 - \frac{365 \times 364 \times 363 \cdots (365 - n + 1)}{365^n}$$

where, in this final form, we have written down the explicit form of the n'th factor in the numerator. (Maybe you had better check up, by adopting some reasonably small value for n, that the n'th factor is really $365 - n + 1$ and not $365 - n$, which you might have wrongly guessed.)

Having worked out this neat general formula, let's look at the numerical results it gives for a few interesting values of n.

When there are 10 persons in a room together, this formula shows that the probability is 0.117 (or better than one chance in nine) that at least two of them have the same birthday. For $n = 22$ the formula gives $p = .476$; whereas for $n = 23$ it gives $p = 0.507$. So if there are 23 persons in a room, there is better than an even chance that at least two have the same birthday.

Most people find this surprising. But even more surprising is the fact that with 50 persons, the probability is 0.970. And with 100 persons, the odds are better than three million to one that at least two have the same birthday.

In World War II, I mentioned these facts at a dinner attended by a group of high-ranking officers of the Army and Navy. Most of them thought it incredible that there was an even chance with only 22 or 23 persons. Noticing that there were exactly 22 at the table, someone proposed we run a test. We got all the way around the table without a duplicate birthday. At which point a waitress remarked, "Excuse me. But I am the 23rd person in the room, and my birthday is May 17, just like the General's over there." I admit that this story is almost too good to be true (for, after all, the test should succeed only half of the time when the odds are even); but you can take my word for it.

The Un-birthday Present

by Lewis Carroll

*There is one day of the year which is your birthday, but there are
364 days which are not your birthday. Lewis Carroll has some fun
with this fact in the next selection, taken from his book,*
Through the Looking Glass.

*Lewis Carroll is the pen-name of Charles Lutwidge
Dodgson, a mathematician who amused himself and his readers by
playing with ideas taken from mathematics. He was born in 1832,
was graduated from Christ Church, Oxford, in 1854, and was a
lecturer in mathematics there until 1881. He died in 1898.*

"They gave it me," Humpty Dumpty continued thoughtfully as he
crossed one knee over the other and clasped his hands round it,
"they gave it me — for an un-birthday present."

"I beg your pardon?" Alice said with a puzzled air.

"I'm not offended," said Humpty Dumpty.

"I mean, what *is* an un-birthday present?"

"A present given when it isn't your birthday, of course."

Alice considered a little. "I like birthday presents best,"
she said at last.

"You don't know what you're talking about!" cried
Humpty Dumpty. "How many days are there in a year?"

"Three hundred and sixty-five," said Alice.

"And how many birthdays have you?"

"One."

"And if you take one from three hundred and sixty-five
what remains?"

"Three hundred and sixty-four, of course."

Humpty Dumpty looked doubtful. "I'd rather see that
done on paper," he said.

23

Alice couldn't help smiling as she took out her memorandum-book, and worked the sum for him:

$$365$$
$$\underline{1}$$
$$364$$

Humpty Dumpty took the book and looked at it carefully. "That seems to be done right ——" he began.

"You're holding it upside down!" Alice interrupted.

"To be sure I was!" Humpty Dumpty said gaily as she turned it round for him. "I thought it looked a little queer. As I was saying, that *seems* to be done right — though I haven't time to look it over thoroughly just now — and that shows that there are three hundred and sixty-four days when you might get un-birthday presents ——"

"Certainly," said Alice.

"And only *one* for birthday presents, you know. There's glory for you!"

"I don't know what you mean by 'glory,' " Alice said.

Humpty Dumpty smiled contemptuously. "Of course you don't — till I tell you. I meant 'there's a nice knock-down argument for you!' "

"But 'glory' doesn't mean 'a nice knock-down argument,' " Alice objected.

"When *I* use a word," Humpty Dumpty said, in rather a scornful tone, "it means just what I choose it to mean — neither more nor less."

"The question is," said Alice, "whether you *can* make words mean so many different things."

"The question is," said Humpty Dumpty, "which is to be master — that's all."

[Hidden in this humorous conversation between Alice and Humpty Dumpty are two important ideas about the way in which we use words in everyday conversations and in science. 1) In everyday conversations, it is possible for the same words to have entirely different meanings. This leads to confusion when the speaker has one meaning in mind, but the listener has another. For example, when Alice said, "I beg your pardon?" she meant, "I am sorry, but I don't understand what you mean by an un-birthday present." But Humpty Dumpty thought she meant, "I am sorry. I didn't mean to offend you." 2) Such confusion of meanings cannot be tolerated in science, which tries to be clear and precise in all its statements. So, to avoid such confusion, scientists separate certain key words from their ordinary everyday meanings, and give them new meanings that they choose for them. That is, they *define* the technical words that they use, so that they will have only one meaning for all people. The scientist says with Humpty Dumpty, "When *I* use a word, it means just what I choose it to mean — neither more nor less."]

The Hare and the Tortoise—and Other Paradoxes

by James Fey

Columbia University Teachers College

The calculated solution to the birthday problem is surprising to most people because it contradicts what they think of as "common sense." Any statement that seems to be true while contradicting common sense is called a paradox. *In the birthday problem, common sense turns out to be wrong, and the calculated solution turns out to be right. However, there are some paradoxes in which common sense turns out to be right, and the calculated solution turns out to be wrong. One of these is Zeno's famous paradox of the hare and the tortoise, first stated about 450 B.C., and not really understood until about 2400 years later. This paradox and others are described in the following article taken from the May 1968 issue of* The Mathematics Student Journal.

Nearly everyone is familiar with the parable of the hare and the tortoise — one of the oldest presentations of the moral "slow and steady wins the race." Challenged to a race by the faster moving hare, the tortoise actually wins the race by maintaining a slow but methodical pace while the over-confident hare alternates between short bursts of running and longer breaks for sleeping. Clearly the hare could have won the race easily if he had only kept up a steadier pace — or could he?

Certainly in any race between a hare and a tortoise the tortoise should be given some sort of head start advantage. As the tortoise moves along the race course, the hare moves along behind him, probably at a faster pace.

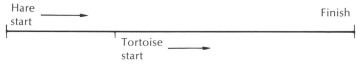

26

In order to pass the tortoise the hare must first pass through the tortoise's starting point. However, when the hare reaches this point, the tortoise has already departed and reached some new point farther along the course.

The hare speeds ahead to the new point, but upon arriving finds that the tortoise has again moved and is at some point still farther along the course.

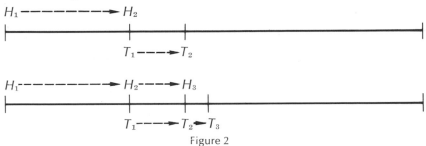

Figure 2

The hare heads for this new point, but when he arrives you can guess what has happened. *No matter how fast the hare runs, when he reaches a point on the course previously occupied by the tortoise, the plodding tortoise has already moved ahead.* The contest is somewhat like a race between horses on a merry-go-round.

The preceding argument seems to prove that any tortoise which is given a headstart and moves steadily can defeat any hare in a race of any distance. In fact, it seems to prove that any creature (man or beast) given the same advantage can win a race against the fastest opposition. However, it is doubtful that such an argument would win blue ribbons at a track meet.

What is the error of reasoning that leads to such a paradoxical conclusion? You might feel that there is a relation between headstart and length of race that allows the hare to win. But re-reading the argument you will find that no matter how short the headstart, the hare keeps running to points the tortoise has already left. Another explanation might be that the hare skips some points and at some stage vaults over the tortoise. But while the hare's feet may not touch each point, his body must certainly pass over each point at some instant in the race.

The argument used to prove that the tortoise will always defeat the hare is a slight variation of one proposed in 450 B.C. by Zeno, a Greek philosopher. This clever Greek went even farther; he proved that no matter how short the race, even the tortoise would *never* reach the finish line! His argument goes as follows: Before the tortoise can reach the finish line, he must reach the halfway point — a trip which takes him a certain finite length of time. But arriving at the halfway point he must next travel to a point halfway to the finish, then halfway again, and again, etc. His trip consists of an infinite number of shorter trips, each of which takes some finite amount of time. *But*, Zeno reasoned, *any sum of an infinite number*

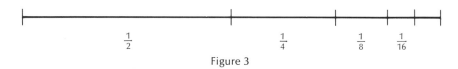

Figure 3

of time intervals must be infinite. Therefore the tortoise can never reach the finish line in a finite amount of time.

You may react to these strange arguments with the comment, "Very clever, but what do they have to do with the real mathematics?" Since common mathematical terms such as *distance, point, motion,* and *infinity* appear at crucial points in each paradox, the results have long generated philosophical debate among mathematicians. If our intuitive knowledge about points and numbers leads to logical conclusions that are contrary to experience, it is possible that other applications of mathematical principles might produce dangerously false results.

Although the best scientists, mathematicians and philosophers in the centuries after Zeno tried to explain away his strange conclusions, little important progress was made until Galileo proposed two more paradoxes in 1636. *The key to Zeno's paradoxes is proper understanding of the infinite;* and with his new (but also paradoxical) findings, Galileo brought fresh insight to the problem. He observed that while not every positive integer is the square of another integer (e.g., there is no integar whose square is 5), each positive integer can be matched with an integer that is a perfect square.

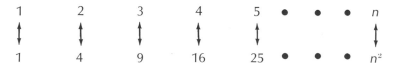

Although $\{1, 4, 9, 16, 25, \cdots \}$ seems to have fewer elements than

28

{1, 2, 3, 4, 5, ⋯ }, it is possible to set up a one-to-one correspondence between the two sets. The perfect squares won't run out! **With finite sets related in a similar way, no such correspondence can be found;** try to match {1, 2, 3, 4} and {1, 2, 3, 4, 5} in a one-to-one manner. *Strange things happen in the world of the infinite.*

Even more important for the plight of the hare racing a tortoise, Galileo also noted that the lengths of two line segments cannot be compared by counting the number of points in the segments. There are exactly as many points in a three inch segment as in a two inch segment. The points can clearly be matched as follows:

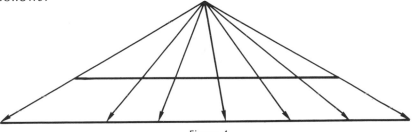

Figure 4

With the recognition that segments of differing lengths can have the same number of points, the paradox of the hare and the tortoise can be explained satisfactorily. While it is true that the hare must touch as many points as the tortoise, this does not prevent him from covering a greater distance (even though the points on the hare's path are no bigger than those on the path of the tortoise). Lack of ambition could still prove the downfall of the racing hare, but he has nothing to fear from Zeno.

While Galileo's insight into the infinite provides an explanation for Zeno's paradox, the explanation itself contains a paradox. One of the sacred principles of Greek mathematics was the belief that the *whole is always greater than any of its parts.* Yet Galileo showed that a *part* of the positive integers, {1, 4, 9, 16, ⋯ }, has just as many elements as the *whole* set of positive integers. He explained this apparent contradiction by saying that both sets are infinite and the attributes *equal, greater,* and *less* are not applicable to infinite quantities. It was not until the end of the 19th century that Galileo's paradox was fully understood.

Then Georg Cantor, founder of the theory of sets, proposed that two sets be assigned the same number (finite or infinite) if they could be matched element by element. For example, {1, 2, 3} has the same number of elements as $\{a, b, c\}$, the set of states in the United States has the same number of elements as the set of state governors, or (in a case involving infinite sets) the set of positive integers has the same number of elements as the set of even positive integers.

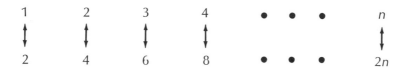

No sooner had Cantor laid down his theory of sets and cardinal numbers than new paradoxes appeared to challenge his ideas. According to his definition, many sets have the same number of elements as some of their subsets. For example, $\{1, 2, 3, 4, 5, \cdots\}$ has the same number of elements as $\{2, 4, 6, 8, 10, \cdots\}$ and $\{3, 6, 9, 12, 15, \cdots\}$ and $\{15, 30, 45, 60, 75, \cdots\}$. In fact, Cantor said a set is infinite if and only if it can be placed in one-to-one correspondence with one of its proper subsets. *He argued that the principle of a whole being greater than any of its parts is true only for finite sets.* This denial of a previously unquestioned fact of mathematical life sparked a furious controversy among 19th century mathematicians. To add fuel to this controversy, Cantor also found some infinite sets that have more elements than the set of integers. He showed, for example, that there is no way to match the integers one-to-one to the set of real numbers. **In other words, the set of**

✻ **real numbers is of a higher order of infinity than the set of integers. Cantor proved, moreover, that there are an infinite number of infinite numbers!**

> ✻ **The set of real numbers is the set of all non-terminating decimal fractions, such as**
> **2.500000...,**
> **9.13131313...,**
> **0.1010010001...,**
> **etc.**
> **including those that repeat and those that do not repeat. (See page 155.)**

Just as Cantor's theory showed promise of introducing some order into the mysterious world of the infinite, mathematicians who were suspicious of the entire business involving sets and infinity were encouraged by the discovery of new paradoxes centering around the use of sets. Bertrand Russell posed the following riddle: The village barber shaves all people in the village who do not shave themselves. Who shaves the barber? Does he belong to the set of people who shave themselves (and thus are not shaved by the barber) or is he shaved by the village barber (and thus by himself)? Evidently care must be exercised when working with this notion of set. Cantor actually offered a kind of contradiction to his own theory when he proved that there is a set which is bigger than the biggest possible set!

These and other paradoxical consequences of the theory of sets made mathematicians very cautious in their use of set ideas. Just as Zeno's paradoxes sparked philosophical investigation into the notions of point, distance, and infinity, the newer paradoxes provoked serious examination of other previously unquestioned mathematical facts and methods. Mathematicians of the 20th century have worked on and, to some extent, solved many paradoxes of set theory and the infinite. However, no sooner is one family of paradoxes resolved than a new batch of strange riddles appears.

If you would like to try your hand at explaining an important paradox, study this one attributed to Zeno (who had an active but pesky mind): An arrow shot from a bow is apparently in continuous motion from the time it leaves the bow until it reaches its destination. However, at any given instant of time the arrow is at exactly one point or location in space. In order to be *at a point*, the arrow must at that instant be at rest. Since the arrow is at a point in each instant of its flight and since being at a point means being at rest, the arrow must be at rest throughout its flight!

Splitting the Ring

from *Magic House of Numbers*
by Irving Adler

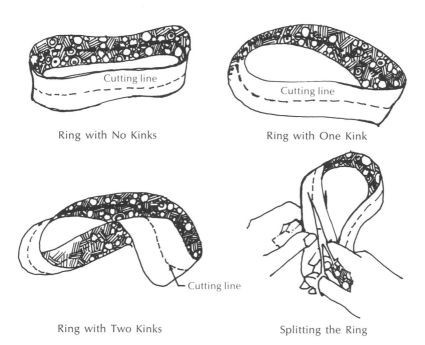

Ring with No Kinks   Ring with One Kink

Ring with Two Kinks   Splitting the Ring

Unfold a double page of regular size newspaper and cut from the longest edge three strips of paper, each one inch wide. Bring the ends of one of the strips together to form a ring. Make sure that there are no kinks in the ring, and then fasten the ends with glue or with adhesive tape. Make a ring with the second strip, but twist one of the ends over once before fastening the ends together. Make a ring with the third piece, but give it two twists before fastening the ends together. Now, with your scissors, make a cut halfway between the edges of the first ring, and extend the cut all the way around the ring. When you finish cutting, you will have two rings that can be separated.

Now make the same kind of cut around each of the other rings. What will you get as a result? Try it and get a big surprise.

A One-sided Surface

The ring with no kinks, described in the preceding selection, is like the surface of a tube. It has two sides, an *inside* and an *outside*. If you grasp the ring between your thumb and forefinger so that your thumb touches the outside of the ring, then your forefinger touches the inside of the ring. If there were two ants crawling around on the ring, one on the inside, and the other on the outside, without ever crossing over an edge of the ring, then neither ant could ever get to where the other one is.

The situation is quite different on the ring with one kink. Here, if there are two ants standing on what seem to be opposite sides, one can crawl to the other without crossing over an edge. This is shown clearly in the remarkable drawing on page 34, reproduced from the book, *The Graphic Work of M. C. Escher*. The ring with one kink is known as a *Möbius strip*. The fact that an ant can crawl to any part of the surface of a Möbius strip without crossing over an edge shows that what seem to be the inside and the outside of the strip are not really separate sides. They are part of the same side. *A Möbius strip is a surface that has only one side!* How many edges does a ring with no kinks have? How many edges does a Möbius strip have?

M. C. Escher is a Dutch artist. His drawings of space paradoxes have delighted people all over the world for many years.

On page 34 is another drawing of a paradox by M. C. Escher. We see in the picture a waterfall in which water drops from a higher level to a lower level. Then it flows in what seems to be a horizontal channel. But after making four right-angled turns as it moves along the channel, the water ends up on the upper level again, ready to fall once more! Mr. Escher has the water turning a waterwheel, so the endless flow of water produces an endless turn-

33

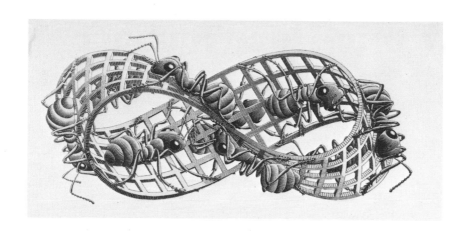

ing of the wheel. If such a waterfall-wheel combination could really be made, it would be a perpetual motion machine.

The paradox in this case is the result of a trick Mr. Escher used in making the drawing. What is the trick? The answer to this question is printed below upside down.

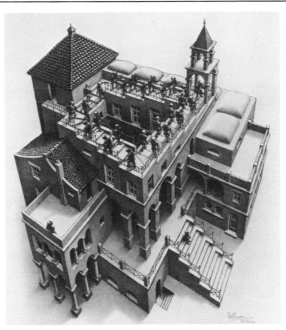

In the above paradoxical drawing also by M. C. Escher, there are two files of men walking on the roof of a building around a central court. The roof consists of a series of steps. Each man in the outer file, walking clockwise, is constantly climbing the steps. But in spite of the climb, he finds himself, after one trip around the court, on the same level from which he started. Meanwhile, each man in the inner file, walking counterclockwise, is constantly going down the steps. But in spite of his descent, he also finds himself, after one trip around the court, on the same level from which he started. Since the constant climbing or descending doesn't get them anywhere, let us hope that at least they like the exercise!

(The water-channel zigzags across the page. There are three places where Mr. Escher drew posts between a zig and a zag, thus making them seem to be on different levels.)

Rates, Paradoxes and Fallacies

by Irving Adler

There are some interesting and amusing number paradoxes connected with the idea of rate. They are described in the next selection, taken from A New Look at Arithmetic *by Irving Adler.*

A RATE FALLACY

A common error that people make is to confuse an average speed with the average of two numbers. Suppose, for example, we consider this problem: An automobile went from one town to another at a speed of 20 miles per hour, and returned at a speed of 30 miles per hour. What is the average speed for the round trip? Most people will answer this equation by thinking "The average of 20 and 30 is $\frac{1}{2}(20 + 30) = 25$. So the average speed for the round trip is 25 miles per hour." However, this answer is wrong. The correct way to compute the average speed for the round trip is to use the formula

$$\text{average speed} = \frac{\text{total distance}}{\text{total time}}$$

To permit us to solve the problem without using algebra, let us assume that the distance between the two towns is 60 miles. (The answer will be the same no matter what distance we assume.) Then the total distance for the round trip is 120 miles. The first part of the trip, covering 60 miles at a speed of 20 miles per hour takes 3 hours. The return trip, covering 60 miles at a speed of 30 miles per hour, takes 2 hours. Then the total time for the round trip is 5 hours. Consequently, the average speed is $\frac{120}{5}$ miles per hour, or 24 miles per hour.

A RATE PARADOX

There is another problem involving an average rate, in which it is very easy to be trapped into making the same error. Here the wrong

36

answer seems so beguilingly simple and plausible that the correct answer appears hard to believe at first.

Problem: If an automobile goes from New York to Chicago at a speed of 30 miles per hour, how fast should it return to make the average speed for the round trip 60 miles per hour? There is a strong temptation to answer by saying, "Return at 90 miles per hour, since the average of 30 and 90 is $\frac{1}{2}(30 + 90) = 60$." However, as we have already seen, this is not the correct way to compute an average speed. To make the computation easily without using algebra, let us assume that the distance from New York to Chicago is 900 miles. (The answer will be the same no matter what distance we assume.) The trip from New York to Chicago is a 900-mile trip, done at a speed of 30 miles per hour. Then this part of the trip takes $\frac{900}{30}$ hours, or 30 hours. The round trip is an 1800 mile trip. If it is to be done at an average speed of 60 miles an hour, then the round trip should take $\frac{1800}{60}$ hours, or 30 hours. But 30 hours have already been used up in the first leg of the journey. Therefore, the return trip must be made in 0 hours, which is obviously impossible. Thus, there is no return trip speed that is large enough to make the average speed for the round trip 60 miles per hour.

A PERCENT FALLACY

A common error that people make when they use rates expressed as percents is to assume that successive rates of increase or decrease are additive. For example, they assume that two successive increases of 10% are equivalent to a single increase of 20%; or that two successive decreases of 10% are equivalent to a single decrease of 20%; or that an increase of 10% followed by a decrease of 10% is the same as no change at all. To expose these fallacies, let us examine each of these situations separately.

Two successive increases of 10%. Let x be any number which is subjected to two successive increases of 10%. After the first increase, the number is changed to 110% of its original value. That is, x is replaced by 1.10x. When this number is increased by 10%, it too is changed to 110% of its original value. That is, 1.10x is replaced by 1.10 (1.10)x, or 1.21x. The net result is that x has been replaced by 1.21x. That is, there has been a 21% increase. *Two successive increases of 10% are equivalent to a single increase of 21%.*

Two successive decreases of 10%. When the number x is decreased by 10%, it is changed to 90% of its original value. That is, x is replaced by .90x. When this number is decreased by 10%, it too is changed to 90% of its original value. That is, .90x is replaced by .90 (.90x), or .81x. The net result is that x has been replaced by .81x. That is, there has been a 19% decrease. *Two successive decreases of 10% are equivalent to a single decrease of 19%.*

An increase of 10% followed by a decrease of 10%. When x is increased by 10% it is replaced by 1.10x. When 1.10x is decreased by 10%, it is replaced by .90 (1.10x), or .99x. The net result is that x has been replaced by .99x. That is, there has been a decrease of 1%. *An increase of 10% followed by a decrease of 10% is equivalent to a single decrease of 1%.*

EXERCISES

1. Change 50 miles per hour into a number of feet per second.
2. If a man walks to town at a speed of 3 miles per hour, and returns by bicycle at a speed of 10 miles per hour, what is his average speed for the round trip?
3. What single rate of increase is equivalent to two successive increases of 20%?
4. What single rate of decrease is equivalent to two successive decreases of 20%?
5. What is the net effect of an increase of 20% followed by a decrease of 20%?
6. A man bought and then sold two articles, selling each for $100. One sale resulted in a loss of 10% of the purchase price. The other sale resulted in a gain of 10% of the purchase price. Altogether did he gain, lose or break even?

(See page 50 for the answers to these exercises.)

Answers to exercise on page 45.

1974	214,406,000	2,787,000	217,193,000
1975	217,193,000	2,823,000	220,016,000
1976	220,016,000	2,860,000	222,876,000
1977	222,876,000	2,897,000	225,773,000
1978	225,773,000	2,935,000	228,708,000
1979	228,708,000	2,973,000	231,681,000
1980	231,681,000	3,012,000	234,693,000

PART 3

Growth

The Mathematics of Growth

There are many different kinds of growth that are possible. The simplest kind takes place through the addition of equal amounts at equal intervals of time. An example of this kind of growth is saving without interest by making regular deposits in a piggy bank. If you put 5¢ in the bank each week, the total amount in the bank grows by 5¢ each week, and no more.

Another type of growth takes place when the amount added to the growing sum during one period shares in the growth during the next period. An example of this kind of growth is the growth of a bank deposit earning compound interest. If an initial deposit of $100 earns interest at the rate of 4% per year, compounded annually, then at the end of the first year the deposit grows by $4 (4% of $100); at the end of the second year it grows by $4.16 (4% of $104); at the end of the third year it grows by $4.33 (4% of $108.16); and so on. Although the rate of growth is the same from year to year, namely 4%, the amount of growth increases from year to year.

Population growth is like the growth of a deposit earning compound interest.

The selections that follow are all related to important and interesting questions concerning growth or the mathematics of growth. They deal with the true cost of borrowing money; the explosion of the human population; and some interesting forms of growth in nature.

The Total Interest Paid on a Mortgage Loan

Few people realize that the interest alone that they pay on a mortgage may be more than the amount of money that they borrowed. The two tables that follow show the calculations necessary to figure out the total interest paid on mortgages that run about 20 years and about 30 years respectively, when the interest rate is 8%. To simplify the calculations, we assume that only one payment is made a year, and all sums are rounded off to the nearest dollar. In the first table, a $10,000 loan is repaid by 20 annual payments of $1000 and a final payment of $917. In the second table, a $10,000 loan is repaid by 28 annual payments of $900 and a final payment of $500. The total amount paid under the 21 year mortgage is $20,917 of which $10,917 is interest. The total amount paid under the 29 year mortgage is $25,700 of which $15,700 is interest. This means that you pay almost $5000 more for the privilege of taking eight more years to pay.

Each of these tables is calculated in this way: Column (4) is 8% of column (2). Column (5) = column (3) — column (4). Column (6) = column (2) — column (5). In each line after the first year, the number in column (2) is the same as the number in column (6) in the preceding line.

$10,000 Loan at 8%
Repaid in 21 years

Age of loan in years (1)	Unpaid balance from previous year (2)	Payment made (3)	Interest at 8% (4)	Reduction of mortgage (5)	Balance owed (6)
1	10,000	1000	800	200	9,800
2	9,800	1000	784	216	9,584
3	9,584	1000	767	233	9,351
4	9,351	1000	748	252	9,099
5	9,099	1000	728	272	8,827
6	8,827	1000	706	294	8,533
7	8,533	1000	683	317	8,216
8	8,216	1000	657	343	7,873
9	7,873	1000	630	370	7,503
10	7,503	1000	600	400	7,103
11	7,103	1000	568	432	6,671
12	6,671	1000	534	466	6,205
13	6,205	1000	496	504	5,701
14	5,701	1000	456	544	5,157
15	5,157	1000	413	587	4,570
16	4,570	1000	366	634	3,936
17	3,936	1000	315	685	3,251
18	3,251	1000	260	740	2,511
19	2,511	1000	201	799	1,712
20	1,712	1000	137	863	849
21	849	917	68	849	0
	Total	20,917	10,917	10,000	

$10,000 Loan at 8%
Repaid in 29 Years

Age of loan in years	Unpaid balance from previous year	Payment made	Interest at 8%	Reduction of mortgage	Balance owed
(1)	(2)	(3)	(4)	(5)	(6)
1	10,000	900	800	100	9,900
2	9,900	900	792	108	9,792
3	9,792	900	783	117	9,675
4	9,675	900	774	126	9,549
5	9,549	900	764	136	9,413
6	9,413	900	753	147	9,266
7	9,266	900	741	159	9,107
8	9,107	900	729	171	8,936
9	8,936	900	715	185	8,751
10	8,751	900	700	200	8,551
11	8,551	900	684	216	8,335
12	8,335	900	667	233	8,102
13	8,102	900	648	252	7,850
14	7,850	900	628	272	7,578
15	7,578	900	606	294	7,284
16	7,284	900	583	317	6,967
17	6,967	900	557	343	6,624
18	6,624	900	530	370	6,254
19	6,254	900	500	400	5,854
20	5,854	900	468	432	5,422
21	5,422	900	434	466	4,956
22	4,956	900	396	504	4,452
23	4,452	900	356	544	3,908
24	3,908	900	313	587	3,321
25	3,321	900	266	634	2,687
26	2,687	900	215	685	2,002
27	2,002	900	160	740	1,262
28	1,262	900	101	799	463
29	463	500	37	463	0
	Total	25,700	15,700	10,000	

The Population of the United States

The population of the United States in 1970, as reported by the Bureau of the Census, was about 204 million. This was an increase of 14.2% over the population in 1960. In the preceding decade, from 1950 to 1960, the population had increased by 18.5%.

On September 1, 1970 at 3 P.M., the population of the United States was 205,730,682, and it was increasing at the rate of one person every $15\frac{1}{2}$ seconds.

The 14.2% increase in population between 1960 and 1970 represents an average annual rate of increase of about 1.3%. If we asume that this rate of increase will continue unchanged into the future, we can predict what the population will be for each year after 1970 by making a simple calculation like the computation of compound interest. In the table below, the results of the computation are shown for the first few years. Be your own census forecaster by completing the computation for the remaining years listed in the table. To make the computation, remember a) For each year after 1970, the number in column (2) is the same as the number in column (4) for the preceding year; b) column (3) is obtained by multiplying column (2) by .013; c) column (4) is obtained by adding column (3) to column (2). Round off all numbers in columns (2), (3), and (4) to the nearest thousand.

Predicted Population of the United States
(if the annual rate of increase is 1.3%)

Year (1)	Population at start of year (2)	Growth during the year (3)	Population at end of year (4)
1970	204,000,000	2,652,000	206,652,000
1971	206,652,000	2,686,000	208,938,000
1972	208,938,000	2,716,000	211,654,000
1973	211,654,000	2,752,000	214,406,000
1974			
1975			
1976			
1977	Answers are given on page 38		
1978			
1979			
1980			

The next table shows what the population of the United States will be in future decades if it continues to increase at the rate of about 14% every ten years.

Predicted Population of the United States
(if the rate of increase is 14% every ten years)

Year (1)	Population in millions at start of decade (2)	Growth in millions during decade (3)	Population in millions at end of decade (4)
1970	204	29	233
1980	233	33	266
1990	266	37	303
2000	303	42	345
2010	345	48	393
2020	393	55	448

Notice that an increase of 14% every ten years almost doubles the population in fifty years.

New Cities for America

by Edgardo Contini

*While the population of the United States is now growing at a rate
of about 14% in ten years, the population of its cities is growing
at a faster rate. This growth is creating serious problems that
will have to be solved in coming decades. Some of these problems
are described in the next selection, part of an article that appeared
in the October-November 1967 issue of* The Center Magazine.*

"In the next thirty-five years we must literally build a second America
— putting in as many houses, schools, apartments, parks and offices
as we have built through all the time since the Pilgrims arrived on
these shores."

— President Lyndon B. Johnson

This statement puts into focus the enormity of the task before us
better than any set of statistics. Yet it may be worthwhile to refer to
a set of figures derived from U.S. Census statistics applicable to
metropolitan areas. They summarize past conditions and projections
of population growth and its distribution between rural and urban
locations.

**The tabulation tells us that between the present and the
turn of the century urban population in the U.S. will be more than
double what it is now, and will represent seventy-five percent of the
total population.**

All present trends and predictions indicate that without a
serious national effort to encourage other patterns of growth, most
of the increase will occur around existing metropolitan complexes.

*Growth of the city contributes increasing opportunities
for employment, residential choice, education, recreation and cul-*

*Reprinted, with permission, from the October 1967 issue of *The Center
Magazine,* a publication of the Center for the Study of Democratic Institutions in
Santa Barbara, California.

ture. But growth also aggravates the problems of the community, increases the complexity of transportation and service systems, and generally requires more for the support of public facilities and services.

As growth continues indefinitely, there is a point of diminishing return. Benefits tend to decline while problems and liabilities increase. There may be disagreement about exactly where this point occurs, but *when urban settlement reaches the order of ten or twenty millions, the residents of Megalopolis pay a high price for the dubious advantages of bigness.* The only residual benefit is an increase in business or employment opportunities. Yet it can be argued that the average level of employment results from the over-all rate of national population growth and productivity, not from the size of population in any specific area. As for increased cultural and educational opportunities, beyond a certain point they fail to follow urban growth. When Megalopolis requires duplication of museums, theaters, civic centers, and multiple urban centers, each center remains limited in its influence. As Megalopolis continues to grow its residents limit themselves to their immediate neighborhood and actually enjoy fewer cultural and recreational advantages than residents of a smaller city.

Year	U.S. Population	Population Inside S.M.S.A.*	% of Total	Outside S.M.S.A. Population	% of Total
1940	132,165,000	72,834,000	55.1	59,331,000	44.9
1950	151,326,000	89,317,000	59.0	62,009,000	41.0
1960	179,323,000	112,885,000	63.0	66,438,000	37.0
1965	192,185,000	123,813,000	64.4	68,372,000	35.6
1970	208,249,000	137,444,000	66.0	70,805,000	34.0 ✷
1980	244,566,000	168,751,000	69.0	75,815,000	31.0
1990	287,472,000	206,980,000	72.0	80,492,000	28.0
2000	337,472,000	253,104,000	75.0	84,368,000	25.0

*Standard Metropolitan Statistical Areas

✷ **The figure for 1970 U.S. population as projected in this table is 208+ million, whereas the figure reported by the Bureau of the Census for 1970 (page 45) was about 204 million; in other words a 4 million margin of error. This underlines that projected figures are and can only be approximations.**

47

Other disadvantages of Megalopolis are more obvious. The resident is robbed of the benefit of leisure time. Travelling through mile after mile of uninterrupted urban structure to reach the country for a breath of fresh air is a depressing experience, regardless how comfortable the seat on the commuter train.

If the projection of growth cannot be challenged and the trend toward urbanization is not reversible, are there alternatives to the predicted image of the United States of tomorrow — two-thirds of its population concentrated in a dozen clusters of urbanization of fifteen to twenty million each?

The alternative consists of undertaking as a matter of national policy the creation of New Cities — "new" in that they must be viewed as urban organizations responsive to human needs and aspirations; "cities" in that (unlike the "new towns" dependent upon an adjacent metropolis) they must from the beginning be complete urban structures; "New Cities" in that their self-sufficiency, independence of existing political forces and interests will permit experimentation with techniques that are not possible in the existing cities.

The *New Cities* must be large enough to support a full range of cultural and educational facilities. The optimum size will have to be determined by regional characteristics, the local economic base and other factors. Indefinite expansion beyond optimum size should be discouraged. New political and economic devices and incentives should initially encourage and, at later stages, limit their growth.

Arithmetic and Geometric Growth

Two different kinds of growth were described on page 40. One is the growth of the sum of money in a piggy bank if the same amout is added to it each week. If the first deposit was 10¢, and each week is given by the sequence of numbers,

$$10, 15, 20, 25, 30, 35, \ldots$$

A sequence like this, in which you *add* the same number (5 in this case) to each term to get the next term, is called an *arithmetic sequence,* and growth that is described by an arithmetic sequence is called *arithmetic growth.*

The second kind of growth described is the growth of a bank deposit earning compound interest. If a deposit of $100 earns interest at the rate of 4% a year compounded annually, the number of dollars with interest, at the end of the first year is $100 + (.04) 100$. By the distributive law, this is the same as $100 (1 + .04)$ or $100 (1.04)$. In other words, the amount of money on deposit at the beginning of the first year is multiplied by 1.04 by the end of the year. Similarly, the amount of money on deposit at the beginning of the second year is multiplied by 1.04 by the end of the second year, etc. The number of dollars on deposit year by year is given by the sequence of numbers,

$$100, 100(1.04), 100(1.04)^2, 100(1.04)^3, \ldots$$ A sequence like this, in which you *multiply* each term by the same number (1.04 in this case) to get the next term, is called a *geometric sequence,* and growth that is described by a geometric sequence is called *geometric growth.*

A very simple arithmetic sequence is the sequence of counting numbers, $1, 2, 3, 4, 5, \ldots$, in which you add 1 to each term to get the next term.

49

A very simple geometric sequence is the sequence of powers of 2: 1, 2, 4, 8, 16, . . . In this sequence, you multiply each term by 2 to get the next term.

If a population increases at a fixed rate from year to year, like money earning compound interest, its growth is an example of geometric growth.

Which of these two sequences is arithmetic, and which is geometric?
    a. 3, 9, 27, 81, 243, . . .
    b. 2, 4, 6, 8, 10, 12, . . .
    See answer upside down/below.

Answer: a. geometric. b. arithmetic.

World Population and Food

In a pamphlet published in 1798, Thomas Robert Malthus asserted that a population tends to grow like a geometric sequence, but its food supply tends to grow like an arithmetic sequence. He concluded that a population tends to grow faster than its food supply, and unless population growth is checked voluntarily by birth control, it is ultimately checked involuntarily by war, disease, and starvation.

The actual relationship between population and food supply is more complicated than Malthus thought. An account of this relationship, based on the scientific knowledge available to us today, is given in the article that begins on page 53.

Malthus's Pamphlet had a great influence on European thought during the first fifty years after it was published. It even suggested to Charles Darwin the key idea in his theory of evolution: survival of the fittest in the struggle for existence. But during the next hundred years, a revolution in agriculture made the ideas of Malthus seem less and less important. The growing use of farm machinery, fertilizers, and insecticides led to a rapid increase in food production even while the number of farmers was declining. It looked as though scientific agriculture might make it possible for man's food supply to keep pace with his population growth.

However, the picture changed after 1945. A worldwide attack on disease led to a sharp decline in the world death rate. Since the drop in the death rate was not matched by a proportionate drop in the birth rate, the population of the world began to grow rapidly. Now again the ideas of Malthus are considered important. People are more and more concerned about the "population explosion" and the danger of worldwide starvation if world population is allowed to run far ahead of the world food supply. This concern is reflected in the news story and the boxed inserts that follow. The news story by the British news service called *Reuters*, was published in *The Christian Science Monitor* on December 2, 1969. The boxed inserts are taken from an article entitled "Can Man Survive?" by James A. Oliver, published in the March 1970 issue of *The American Way* .

Thomas Robert Malthus (1766-1834)   *Woodcarving by William Ransom*

UN Examines Population Boom

by Reuters

UNITED NATIONS, N.Y. — The United Nations is concerned about a worldwide people boom.

According to a UN report issued here, the total world population will increase in 20 years from 3.2 billion in 1965 to 4.9 billion in 1985.

The report shows that the population of poor nations is increasing at twice the rate of wealthy countries. There is a general increase in life expectancy and a population explosion in urban centers throughout the world.

> ### 234 Babies a Minute
> During the next one minute, 234 babies will be born. At the end of an hour the newborns will have brought about a net increase in the world population of 7,900 individuals. By this time tomorrow, the net population of the world will have been increased by a mass of people equal to the total present population of Salt Lake City — 190,000 individuals.

A conference to discuss the problem of the world's soaring population, particularly in the developing countries, has been taking place in Geneva.

The UN Population Commission here had before it two major reports dealing with the population situation and measures and politics affecting fertility, with particular attention given to national family-planning programs.

The first report says that recently recalculated figures showed the world's total population was likely to grow at an annual rate of between 2 and 2.1 percent between 1965 and 1985.

More developed regions are expected to rise by about 2.4 to 2.5 percent. This means world population would grow by 50 percent in 20 years.

Starvation

Each week at least ten thousand human beings die of starvation. Moreover, of the $3\frac{1}{2}$ billion people inhabiting the earth today, two-thirds are undernourished by medical standards.

PERCENTAGE DECLINES

At the beginning of the century nearly half the world's urban population was contained in Europe. But by 1980 Europe will account for less than one-fifth of the world's urban population, the report says.

While the figures show that "cities of historically unprecedented size are emerging and becoming even more numerous," they also imply that the rate of urban growth in less developed regions, which started with smaller urban populations, has been much faster.

"In large organized empires of antiquity, the urban places rarely contained much more than 5 percent of the entire population," the report says. In 1960, however, one-third of all human beings lived in cities.

The second report dealing mainly with family planning says that the basic assumption underlying planning programs is "that individuals will regulate fertility if the rationality of doing so is properly communicated to them and if they are given the necessary information and means."

Pollution

The twin of our population crisis is our undisciplined technology and the vast quantity of wastes, pollutants, contaminants, and poisons that inundate our environment. In the United States alone we annually dump 350 billion tons of garbage and sewage, one billion tons of mining wastes, 48 billion metal cans, 20 million tons of paper, and 7 million worn out automobiles; and we pour 142 million tons of smoke and noxious fumes into the air we breathe. How long can our environment withstand these assaults without massive retaliation?

World Population

Predicted Growth of World Population per Annum (if the rate of increase is 2% per year)			
Year (1)	Population at start of year (in millions) (2)	Growth during the year (3)	Population at end of year (in millions) (4)
1965	3200	64	3264
1966	3264	65	3329
1967	3329	67	3396
1968	3396	68	3464
1969	3464	69	3533
1970	3533	71	3604
1971	3604	72	3676
1972	3676	74	3750
1973	3750	75	3825
1974	3825	77	3902
1975	3902	78	3980

A growth rate of 2% a year is about the
same as a growth rate of 22% in ten years

Predicted Growth of World Population per Decade (if the rate of increase is 22% every ten years)			
Year (1)	Population (in millions) (2)	Growth during the next 10 years (3)	Population after 10 years (in millions) (4)
1965	3200	704	3904
1975	3904	859	4763
1985	4763	1048	5811
1995	5811	1278	7089
2005	7089		

The Shrinking Mississippi River

by Mark Twain

*The predictions of future population printed on pages 53 and 55
are based on the assumption that the present rate of growth will
continue into the future. While this assumption may be right, it is
also possible that it is wrong. If the assumption is wrong, the
predictions may be wrong, and this applies to any kind of prediction.
The next selection is a humorous reminder of this important fact.
It is taken from* Life on the Mississippi, *by the great American
humorist Mark Twain.*

These dry details are of importance in one particular. They give
me an opportunity of introducing one of the Mississippi's oddest
peculiarities — that of shortening its length from time to time. If
you will throw a long, pliant apple-paring over your shoulder, it will
pretty fairly shape itself into an average section of the Mississippi
River; that is, the nine or ten hundred miles stretching from Cairo,
Ill., southward to New Orleans, the same being wonderfully
crooked, with a brief straight bit here and there at wide intervals.
The two-hundred-mile stretch from Cairo northward to St. Louis is
by no means so crooked, that being a rocky country which the river
cannot cut much.

The water cuts the alluvial banks of the "lower" river into
deep horseshoe curves; so deep, indeed, that in some places if you
were to get ashore at one extremity of the horseshoe and walk
across the neck, half or three-quarters of a mile, you could sit down
and rest a couple of hours while your steamer was coming around
the long elbow at a speed of ten miles an hour to take you on
board again. When the river is rising fast, some scoundrel whose
plantation is back in the country, and therefore of inferior value,
has only to watch his chance, cut a little gutter across the narrow
neck of land some dark night, and turn the water into it, and in a

wonderfully short time a miracle has happened: to wit, the whole Mississippi has taken possession of that little ditch, and placed the countryman's plantation on its bank (quadrupling its value), and that other party's formerly valuable plantation finds itself away out yonder on a big island; the old watercourse around it will soon shoal up, boats cannot approach within ten miles of it, and down goes its value to a fourth of its former worth. Watches are kept on those narrow necks at needful times, and if a man happens to be caught cutting a ditch across them, the chances are all against his ever having another opportunity to cut a ditch.

Pray observe some of the effects of this ditching business. Once there was a neck opposite Port Hudson, La., which was only half a mile across in its narrowest place. You could walk across there in fifteen minutes; but if you made the journey around the cape on a raft, you traveled thirty-five miles to accomplish the same thing. In 1722 the river darted through that neck, deserted its old bed, and thus shortened itself thirty-five miles. In the same way it shortened itself twenty-five miles at Black Hawk Point in 1699. Below Red River Landing, Raccourci cut-off was made (forty or fifty years ago, I think). This shortened the river twenty-eight miles. In our day, if you travel by river from the southernmost of these three cut-offs to the northernmost, you go only seventy miles. To do the same thing a hundred and seventy-six years ago, one had to go a hundred and fifty-eight miles — a shortening of eighty-eight miles in that trifling distance. At some forgotten time in the past, cut-offs were made above Vidalia, La., at Island 92, at Island 84, and at Hale's Point. These shortened the river, in the aggregate, seventy-seven miles.

Since my own day on the Mississippi, cut-offs have been made at Hurricane Island, at Island 100, at Napoleon, Ark., at Walnut Bend, and at Council Bend. These shortened the river, in the aggregate, sixty-seven miles. In my own time a cut-off was made at American Bend, which shortened the river ten miles or more.

Therefore the Mississippi between Cairo and New Orleans was twelve hundred and fifteen miles long one hundred and seventy-six years ago. It was eleven hundred and eighty after the cut-off of 1722. It was one thousand and forty after the American Bend cut-off. It has lost sixty-seven miles since. Consequently, its length is only nine hundred and seventy-three miles at present.

Now, if I wanted to be one of those ponderous scientific people, and "let on" to prove what had occurred in the remote past, by what had occurred in a given time in the recent past, or what will occur in the far future by what has occurred in late years, what an opportunity is here! Geology never had such a chance, nor such exact data to argue from! Nor "development of species," either! Glacial epochs are great things, but they are vague — vague. Please observe:

In the space of one hundred and seventy-six years the Lower Mississippi has shortened itself two hundred and forty-two miles. That is an average of a trifle over one mile and a third per year. Therefore, any calm person, who is not blind or idiotic, can see that in the Old Oolitic Silurian Period, just a million years ago next November, the Lower Mississippi River was upward of one million three hundred thousand miles long, and stuck out over the Gulf of Mexico like a fishing-rod. And by the same token any person can see that seven hundred and forty-two years from now the Lower Mississippi will be only a mile and three-quarters long, and Cairo and New Orleans will have joined their streets together, and be plodding comfortably along under a single mayor and a mutual board of aldermen. There is something fascinating about science. One gets such wholesale returns of conjecture out of such a trifling investment of fact.

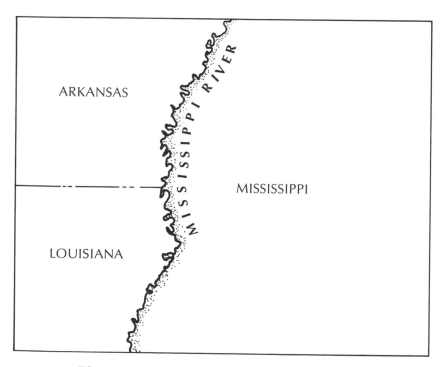

Rabbits and the Fibonacci Numbers

by N. N. Vorob'ev

*Rabbits breed more rapidly than people. If there were no check
on the growth of the rabbit population, there would be a
"rabbit population explosion" that would be far more serious than
the human population explosion that people are now worried
about. Australia, in fact, once experienced a rabbit population
explosion as a result of bringing rabbits into the country without
also bringing in the rabbit's natural enemies.*

*Almost 800 years ago an Italian mathematician studied a
theoretical rabbit population explosion. His study led to a very
interesting sequence of numbers called* Fibonacci numbers, *after the
nickname by which he is known. (The nickname is pronounced
fee-baw-ná-chee.) We are introduced to these numbers in the
next selection, taken from the book about them by the
Russian mathematician N. N. Vorob'ev.*

The ancient world was rich in outstanding mathematicians. Many achievements of ancient mathematics are admired to
this day for the acuteness of mind of their authors, and the names
of Euclid, Archimedes and Hero are known to every educated
person.

Things are different as far as the mathematics of the
Middle Ages is concerned. Apart from Vieta, who lived as late as the
sixteenth century, and mathematicians closer in time to us, a school
course of mathematics does not mention a single name connected
with the Middle Ages. This is, of course, no accident. In that epoch
the science developed extremely slowly, and mathematicians of real
stature were few.

The greater then is the interest of the work *Liber Abacci*
("a book about the abacus"), written by the remarkable Italian
mathematician, Leonardo of Pisa, who is better known by his nick-

name Fibonacci (an abbreviation of *filius Bonacci*). This book, written in 1202, has survived in its second version, belonging to 1228.

Liber Abacci is a voluminous work, containing nearly all the arithmetical and algebraic knowledge of those times. It played a notable part in the development of mathematics in Western Europe in subsequent centuries. In particular, it was from this book that Europeans became acquainted with the Hindu (Arabic) numerals.

The theory contained in *Liber Abacci* is illustrated by a great many examples, which make up a significant part of the book.

Let us consider one of these examples, that which can be found on pages 123-124 of the manuscript of 1228:

"How many pairs of rabbits are born of one pair in a year?" This problem is stated in the form:

"Someone placed a pair of rabbits in a certain place, enclosed on all sides by a wall, to find out how many pairs of rabbits will be born there in the course of one year, it being assumed that every month a pair of rabbits produces another pair, and that rabbits begin to bear young two months after their own birth.

"As the first pair produces issue in the first month, in this month there will be 2 pairs. Of these, one pair, namely the first one, gives birth in the following month, so that in the second month there will be 3 pairs. Of these, 2 pairs will produce issue in the following month, so that in the third month 2 more pairs of rabbits will be born, and the number of pairs of rabbits in that month will reach 5; of which 3 pairs will produce issue in the fourth month, so that the number of pairs of rabbits will then reach 8. Of these, 5 pairs will produce a further 5 pairs, which, added to the 8 pairs, will give 13 pairs in the fifth month. Of these, 5 pairs do not produce issue in that month but the other 8 do, so that in the sixth month 21 pairs result. Adding the 13 pairs that will be born in the seventh month, 34 pairs are obtained: added to the 21 pairs born in the eighth month it becomes 55 pairs in that month: this, added to the 34 pairs born in the ninth month, becomes 89 pairs: and increased again by 55 pairs which are born in the tenth month, makes 144 pairs in that month. Adding the 89 further pairs which are born in the eleventh month, we get 233 pairs, to which we add, lastly, the 144 pairs born in the final month. We thus obtain 377 pairs; this is the number of pairs procreated from the first pair by the end of one year.

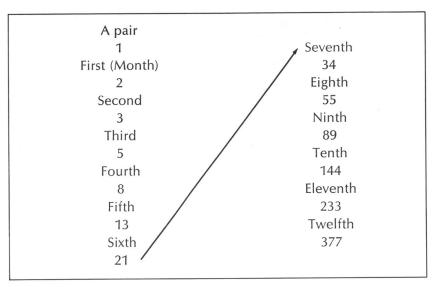

Figure 1

"From [Fig. 1] we see how we arrive at it: we add to the first number the second one i.e. 1 and 2; the second one to the third; the third to the fourth; the fourth to the fifth; and in this way, one after another, until we add together the tenth and the eleventh numbers (i.e. 144 and 233) and obtain the total number of rabbits (i.e. 377); and it is possible to do this in this order for an infinite number of months."

The first thirteen terms of the sequence of numbers obtained in this way are

1, 1, 2, 3, 5, 8, 13, 21, 34, 55, 89, 144, 233 and 377.

They obey this simple rule: *each number starting with 2 is the sum of the last two numbers that come before it in the sequence.* This sequence of numbers is called the Fibonacci sequence to honor the author of the rabbit problem, and its terms are known as Finonacci numbers.

What are the next three Fibonacci numbers after 377?

(Answer: 610, 987, and 1597)

61

The Nature Sequence

by Nathan Ainsworth

The Fibonacci numbers are found very often in nature. We learn about some instances of this in the next selection, reprinted from the February 1970 issue of The Arithmetic Teacher.

Around the beginning of the thirteenth century, Leonardo Fibonacci defined a certain mathematical sequence that proved to have many interesting connections to nature.[1] His sequence has been coined, "The Nature Sequence." Fibonacci's Sequence can be written by starting with two 1's; the next number of the sequence can be formed by adding the two 1's together to get two (2); the next number of the sequence can be found by adding the last two numbers together, which would be $1 + 2 = 3$; other numbers of the sequence can be found in a similar fashion, that is, by adding the last two numbers together. The first few terms of the Fibonacci Sequence are: 1, 1, 2, 3, 5, 8, 13, 21, 34, 55, 89, and so on. Let us now look at some of the interesting occurrences of the Fibonacci Sequence.

THE SUNFLOWER

During the months of July and August, the sunflower is very pretty with its bright yellow face. When the fall months come, the flower petals die, the stamens fall off, and the seeds of the sunflower are exposed. To most people, the sunflower would now be quite unattractive, but with close examination, the beautiful and mysterious arrangement of the sunflower seeds is revealed. When we study the seed arrangement, we notice that the seeds are arranged in such a way that spiral patterns are formed — one spiral pattern proceed-

[1]A sequence is a list of numbers written one after the other so that there is a first number, a second number, a third number, and so on.

62

ing in the clockwise direction and one spiral pattern proceeding in the counterclockwise direction. This is an amazing thing that nature has done, but the most mysterious thing* is that no matter how large or small the sunflower may be, the number of individual clockwise spirals and the number of individual counterclockwise spirals should always be two consecutive numbers of the Fibonacci Sequence. Experiments that the author has carried out have shown a spiral ratio of 55 clockwise and 34 counterclockwise spirals, and in a large sunflower, a spiral ratio of 144 clockwise and 89 counterclockwise spirals. (Figure 1 will help in understanding the spiral arrangements.)

CONES

Many other occurrences in nature have spiral arrangements. One example is in the cones of any evergreen tree. In a cone, whether a pine cone, hemlock cone, spruce cone, or other kinds of pine, there will be a certain number of counterclockwise spirals and a certain number of clockwise spirals. For example, the cone of the white pine tree (fig. 2) has 5 clockwise spirals and 8 counterclockwise spirals. If we compare these two numbers with the Fibonacci Sequence, we can see that 5 and 8 are two consecutive numbers in the sequence.

On the cone of the Loblolly Pine, (Southern Pine) there are 8 counterclockwise and 13 clockwise spirals. The numbers 8 and 13 are again two consecutive numbers of the Fibonacci Sequence.

Fig. 1 Fig. 2

Fig. 1. The seeds of the sunflower are arranged with 34 counterclockwise spirals and 55 clockwise spirals.

Fig. 2. White pine cone has 5 clockwise spirals and 8 counterclockwise spirals. The cross hatching shows 1 clockwise spiral. Two adjacent numbers of Fibonacci's sequence are 5 and 8.

*The 140 year-old puzzle of the occurrence of Fibonacci numbers in plants has recently been solved mathematically by Irving Adler, the editor of this book, so that it is no longer a mystery.

OTHER THINGS IN NATURE

The sections of a pineapple (fig. 3) are arranged in this spiral fashion, having an 8 and 13 spiral arrangement. Most daisies have a 21 and 34 spiral arrangement, again two consecutive numbers of the Fibonacci Sequence. Also, the leaves or buds on a twig make their way down the twig in a spiral fashion. The distance between any two leaves is equal to the sum of the distance between the previous two leaves. This is the exact way in which the Fibonacci Sequence is constructed (see fig. 4). Many twigs will not show this spiral pattern too clearly. For best results, look on the new growth of some bushes that may be near your home. Bushes seem to show the spiral pattern and the spacing better than twigs from a tree.

Another example of this mysterious occurrence is on the cap of an acorn. The spiral rays are rather difficult to count, but nevertheless, they are usually arranged in 21 counterclockwise spirals and 13 clockwise spirals. (See fig. 5.)

The next time you are around trees and flowers, try to notice any spiral patterns that may appear. If you should find some, take several specimens and examine each carefully and count all the spiral rays that go both clockwise and counterclockwise. After recording your findings, compare your results with the Fibonacci Sequence to see if your numbers are, in fact, two consecutive numbers of the Fibonacci Sequence. If you find that your numbers are close to the sequence numbers, then start over and recount the spiral rays.

Fig. 3. Look at this pineapple carefully. Notice the clockwise spirals. The pineapple has an 8 and a 13 spiral arrangement.

64

SHELLS

There are some 80,000 species of shells (snails) or gastropods known today. Gastropods can be identified by their coiling form or their spiral shells. Each of the 80,000 species of shells has a direct connection to the Fibonacci Sequence. The distance between any two spirals is equal to the distance between the next larger spiral distance. For example, if the first spiral distance is 2 cm (centimeters), the second spiral distance is 3 cm, then the third spiral distance, according to the Fibonacci Sequence, would be 5 cm in length. This pattern persists, and the illustrations in figures 6 and 7 show this idea.

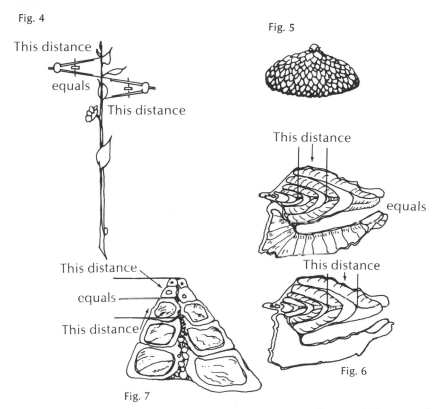

Fig. 4

Fig. 5

This distance

equals

This distance

This distance

equals

This distance

This distance

equals

This distance

This distance

Fig. 6

Fig. 7

Fig. 4. Some twigs from bushes will show the spiral descent of the leaves and buds and will show the most important spacing of the leaves.

Fig. 5. The acorn has 13 clockwise and 21 counterclockwise spirals. These numbers are two adjacent numbers in the Fibonacci sequence.

Fig. 6. The shell illustrated is a cross section of a conch shell. This figure shows the spacing between the spiral sections of the shell.

Fig. 7. A cross section of an Indo-Pacific top shell shows that the distance between two spiral units is equal to the next spiral unit. This is the same way the Fibonacci sequence is constructed.

PART 4

The Very Small

The Atom and Its Nucleus

by Irving Adler

Every material thing is made of small units called atoms. *At the center of each atom is a core called a* nucleus. *The next selection describes how scientists found out what the nucleus is made of. It is taken from the book,* Inside the Nucleus, *by Irving Adler.*

ATOMIC WEIGHTS

The atoms of different chemical elements have different weights. To express these weights chemists assigned a number to each element in the following way: First they assigned the number 16 to oxygen. Then they assigned a number to every other element so that the ratio of the numbers is the ratio of the average weights of the atoms of the elements. For example, since the average sulfur atom is about twice as heavy as the average oxygen atom, they assigned to sulfur a number that is almost 2×16, or 32. Since the average helium atom is about one fourth as heavy as the average oxygen atom, they assigned to helium a number that is almost $\frac{1}{4} \times 16$, or 4. The numbers assigned in this way are called the *atomic weights* of the elements. The atomic weights of all the elements are shown in the table on page 69. Notice that the atomic weight of hydrogen, the lightest of all atoms, is 1.00814, or almost exactly 1.

THE MASS OF AN ATOM

The atoms in the things around us are constantly moving. They are also acted on by various forces which try to change the speed or direction of their motion. Every atom stubbornly resists having the speed or direction of its motion changed. But some atoms resist more than others. Physicists have given a name to the quality of an object that makes it resist changes in its motion. They call it the

68

mass of the object. Experiment shows that the mass of an object is proportional to its weight. Because of this fact, the atomic weight of an atom may also be used as a measure of its mass. So, using 16 as the mass of the average oxygen atom, the mass of the average hydrogen atom is about 1, and the mass of the average helium atom is about 4. The unit in this system of measuring mass is called the *atomic mass unit*. As we see, it is almost exactly equal to the mass of a typical hydrogen atom. Expressed in grams, one atomic mass unit is about 1.7×10^{-24} gram.

THE PERIODIC TABLE

In 1869, the German chemist Meyer and the Russian chemist Mendeleyev, working independently, both made an important discovery. They found that if they listed the chemical elements in about the order of increasing mass, then elements that belonged to the same chemical families occurred at regular intervals in the list. If they broke the list up into rows of elements and put the successive lines under each other in the arrangement shown below, then the elements that came out under one another in the same column were those that had similar chemical properties and belonged to the same chemical family. This special arrangement of the elements is known as the *periodic table* of the elements.

I	II	III	IV	V	VI	VII	VIII			0
H 1 1.008										He 2 4.003
Li 3 6.940	Be 4 9.013	B 5 10.82	C 6 12.011	N 7 14.008	O 8 16.000	F 9 19.00				Ne 10 20.183
Na 11 22.991	Mg 12 24.32	Al 13 26.98	Si 14 28.09	P 15 30.975	S 16 32.066	Cl 17 35.457				Ar 18 39.944
K 19 39.100	Ca 20 40.08	Sc 21 44.96	Ti 22 47.90	V 23 50.95	Cr 24 52.01	Mn 25 54.94	Fe 26 55.85	Co 27 58.94	Ni 28 58.71	
Cu 29 63.54	Zn 30 65.30	Ga 31 69.72	Ge 32 72.60	As 33 74.91	Se 34 70.96	Br 35 79.916				Kr 36 83.80
Rb 37 85.48	Sr 38 87.63	Y 39 88.92	Zr 40 91.22	Nb 41 92.91	Mo 42 95.95	Tc 43 (98)	Ru 44 101.10	Rh 45 102.91	Pd 46 106.4	
Ag 47 107.880	Cd 48 112.41	In 49 114.82	Sn 50 118.70	Sb 51 121.76	Te 52 127.61	I 53 126.91				Xe 54 131.30
Cs 55 132.91	Ba 56 137.36	57-71 La series*	Hf 72 178.50	Ta 73 180.95	W 74 183.86	Re 75 186.22	Os 76 190.2	Ir 77 192.2	Pt 78 195.09	
Au 79 197.0	Hg 80 200.61	Tl 81 204.39	Pb 82 207.21	Bi 83 209.00	Po 84 (210)	At 85 (210)				Rn 86 (222)
Fr 87 (223)	Ra 88 226.05	89-103 Ac series**								

*Lanthanide series:	La 57 138.92	Ce 58 140.13	Pr 59 140.92	Nd 60 144.27	Pm 61 (147)	Sm 62 150.35	Eu 63 152.0	Gd 64 157.26	Tb 65 158.93	Dy 66 162.51	Ho 67 164.94	Er 68 167.27	Tm 69 168.94	Yb 70 173.04	Lu 71 174.99
**Actinide series:	Ac 89 (227)	Th 90 232.05	Pa 91 (231)	U 92 238.07	Np 93 (237)	Pu 94 (242)	Am 95 (243)	Cm 96 (247)	Bk 97 (247)	Cf 98 (251)	Es 99 (254)	Fm 100 (253)	Md 101 (256)	No 102 (254)	Lw 103 (257)

Atomic Symbol — Atomic Number — Atomic Weight

Numbers in parentheses show mass number of most stable known isotope

The periodic table assigns a number to every element, showing its position in the table. This number is called the *atomic number* of the element, and is usually represented by the symbol Z. Thus, for hydrogen, Z = 1. For helium, Z = 2. For uranium, the heaviest atom found in nature, Z = 92.

A BUILDING BLOCK FOR ATOMS?

The atoms of the elements are the building blocks out of which all molecules are made. Is it possible that there is a simple building block out of which the atoms themselves are made? In 1815, William Prout, a London physician, thought he saw an answer to this question in the atomic weights of the elements. The atomic weight of hydrogen is almost exactly equal to 1. The atomic weights of many of the other elements are also almost whole numbers. For example, the atomic weight of carbon is almost exactly 12, the atomic weight of nitrogen is almost exactly 14, and so on. *This gave Prout the idea that the hydrogen atom was the building block out of which all the other atoms were built.* According to this idea a carbon atom was made of 12 hydrogen atoms, and a nitrogen atom was made of 14 hydrogen atoms, and so on.

Prout's idea was popular among chemists for a while. But then it lost favor when it became clear, as atomic weights were measured more accurately, that some of them were very far from being whole numbers. For example, the atomic weight of magnesium is 24.32, and the atomic weight of chlorine is 35.457. However, this obstacle to Prout's theory was swept aside by an important discovery made at the beginning of the twentieth century. It was found that many elements were mixtures of several kinds of atoms that had the same chemical properties but different weights. Atoms that are chemically the same although they have different weights are called *isotopes*. We now know that every element has several isotopes. *When the isotopes of an element are separated and then weighed separately, it is found that each isotope has an atomic weight that is almost a whole number. The fractional atomic weight of a mixture of isotopes is merely the average of the atomic weights of the isotopes.* For example, we know now that magnesium is a mixture of three isotopes. About eighty percent of the mixture has an atomic weight of about 24, about ten percent has an atomic weight of about 25, and the remaining ten percent has an atomic weight of about 26. So the atomic weight of the mixture is about $(.80 \times 24) + (.10 \times 25) + (.10 \times 26) = 24.3$. Similarly, chlorine is a mixture of two isotopes. Seventy-five percent of the mixture has an atomic weight of about 35, and the remaining twenty-five percent has an atomic weight of about 37. So the atomic weight of the mixture is about $(.75 \times 35) + (.25 \times 37) = 35.50$.

The whole number that is nearest to the atomic weight of an isotope of an element is called its *mass number* and is represented by the letter A. Thus, for one isotope of chlorine, $A = 35$, and for the other one, $A = 37$. To name an isotope of an element we write its mass number after its name or we write the mass number as a superscript after its chemical symbol. Thus, the latter isotope of chlorine is called chlorine 37, or Cl^{37}. In the chart on page 72, each shaded square represents an isotope of the element whose name appears at the left.

Since every isotope of an element has an atomic weight that is nearly a whole number, it seems reasonable again to guess that the hydrogen atom, whose atomic weight is 1, is the building block out of which all other atoms are made. We shall see that this guess is almost correct. About 99.95% of the mass of an atom is in its nucleus, and each nucleus is built out of two kinds of building blocks. One of them is a hydrogen nucleus, and the other is a close relative that is capable of becoming a hydrogen nucleus.

ATOMS OF ELECTRICITY

During the second half of the nineteenth century physicists performed many experiments in which they passed an electric current through a closed tube that contained a gas at very low pressure. In one type of experiment, the pressure of the gas was reduced to below one millionth of the pressure of the atmosphere, and a high voltage was connected to two electrodes in the tube. Then a ray could be observed coming from the negative electrode. Sometimes the ray could be seen directly as a bluish thread. Since the negative electrode is called the *cathode*, the ray came to be called a *cathode ray*, and the tube became known as a *cathode ray tube*. A familiar example of a cathode ray tube is the picture tube of a television set. Experiments with cathode rays showed that they are streams of small negatively charged particles which we now call *electrons*. Both the mass and the electrical charges of the electron have since been measured. The mass of the electron is 9.1×10^{-28} gram. The strength of the charge of an electron is 1.601864×10^{-19} coulomb, where the coulomb is the standard unit of charge used in electrical theory. It is customary to represent this amount of charge by the symbol e.

The charge of the electron is written as −e to show that it is a negative charge. In an experiment performed in 1909, R. A. Millikan proved that small electrical charges are always whole number multiples of e. In this way he proved what had long been suspected, that the electron was a kind of atom of negative electricity.

Neutral atom	**Ionized atom**
Each unit of positive charge is balanced by a unit of negative charge	If an electron is removed, the balance is destroyed

In this case,
total charge = +2 −2 = 0

In this case, total
charge = +2 −1 = +1

Electrons are now known to be present in all atoms. A complete atom is electrically neutral because the total negative charge of all the electrons in it is balanced by an equal amount of

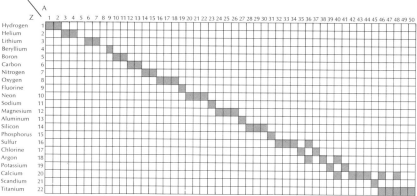

First part of a chart of the stable isotopes of the elements

positive charge. If one or more of the electrons in an atom is removed, the balance is destroyed, and the atom becomes positively charged. Such a positively charged atom is called an *ion*. Ions are produced in some chemical solutions. They are also produced in a gas when fast moving particles collide with atoms of the gas and knock electrons out of them. An ion of special importance is the hydrogen ion, produced by removing one electron from a hydrogen atom. Another name for the hydrogen ion is *proton*. A proton is about 1840 times as heavy as an electron, but it has the same amount of electrical charge. Since the charge of the proton is positive, it is written as +e.

There is evidence that both the electron and proton are spinning like a top. The tendency of a spinning body to keep spinning is called its *angular momentum*. The amount of angular momentum of a spinning body is measured in units known as ergseconds. (The *erg* is a unit of energy.) For particles the size of an atom or smaller, it is convenient to use as the unit of angular momentum an angular momentum of $h/2\pi$, where h has the value 6.625×10^{-27} erg-second, and is known as *Planck's constant, and* π is approximately 3.1416. In terms of this unit, the angular momentum of an electron and a proton are each equal to $\sqrt{\frac{1}{2}(\frac{1}{2}+1)} = \sqrt{\frac{3}{4}} =$ approximately 0.866. For reasons we need not discuss here the number $\frac{1}{2}$ that appears in this formula is called the *spin* of the electron and proton.

DISCOVERY OF THE NUCLEUS

During the nineteenth century experiments with ions, cathode rays, positive rays, and radioactive elements made it clear that there are both positive and negative electrical particles in every atom. This conclusion naturally led to the question, "How are the electrical particles arranged in the atom?" The first step toward the modern answer to this question was taken by Rutherford through an experiment he performed in 1911. Rutherford fired a stream of positively charged particles at a thin sheet of gold foil placed in front of a fluorescent screen. The particles he used were alpha particles coming from the radioactive element radium. Each time a particle passed through the foil and struck the screen, it caused a small flash of light on the screen. The positions of these flashes on the screen showed that most of the alpha particles passed right through the foil, as if nothing were there at all. But some of the particles were turned aside as if something in the foil were pushing them off course. Some of the particles even bounced back from the foil, instead of passing through. The fact that most of the particles went through the foil without meeting any obstacles showed that a large part of each atom in the foil consists of empty space. The detailed pattern in which the rest of the particles were turned aside was analyzed mathematically by Rutherford. The analysis led him to the conclusion that there is a positively charged nucleus at the center of each atom. Alpha particles that passed near the nucleus were turned aside because like electrical charges repel each other.

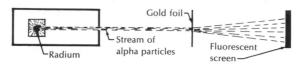

Schematic diagram of the experiment in which Rutherford discovered the nucleus

Rutherford's experiment led to the theory that every atom consists of a positively charged nucleus surrounded by a group of electrons. This experiment and others also showed that if the charge on a proton, $+e$, is taken as a unit, then the charge on the nucleus of an atom is equal to its atomic number Z. When the atom is electrically neutral every unit of positive charge is balanced by a unit of electrical charge supplied by an electron. So an atom whose atomic number is Z has Z electrons surrounding its nucleus. Thus, a hydrogen atom, whose atomic number is 1, consists of a nucleus with charge $+1$, and a single electron revolving around it. A helium atom, whose atomic number is 2, consists of a nucleus with charge $+2$, and two electrons revolving around it. A uranium atom, whose atomic number is 92, consists of a nucleus with charge $+92$, and 92 electrons revolving around it.

How alpha particles are deflected by a nucleus

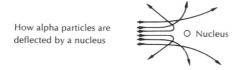

O Nucleus

This discovery showed the real meaning of the order in which the elements are listed in the periodic table. They are listed in the order of the charges on their nuclei, and not in the order of their atomic weights. Isotopes of the same element have different atomic weights, but they have the same nuclear charge.

THE PROTON-ELECTRON THEORY OF THE NUCLEUS

What are the building blocks out of which a nucleus is made? The first important attempt to answer this question was based on Prout's theory that all atoms are made out of hydrogen. Expressed in terms of nuclei, this theory would say that the proton is the building block out of which all nuclei are made. Let us see whether or not this theory fits the facts.

Any building-block theory of the nucleus must be consistent with two facts about the nucleus, that its mass number is A and its nuclear charge number is Z. If we picture a nucleus as being built out of protons only, the number of protons would have to be equal to the mass number of the nucleus. But then, for all nuclei except hydrogen 1 the nuclear charge would come out wrong. As an example, consider the nucleus of carbon 14, for which $Z = 6$ and $A = 14$. To match the mass of this nucleus it would be necessary to combine 14 protons. But then the total charge would be 14, which does not fit the fact that $Z = 6$. Therefore *the heavier nuclei cannot be built out of protons alone.*

To make the theory fit the facts better, it is necessary to introduce another building block in addition to the proton. The first choice made for this second building block was the electron. It was a natural choice to make since electrons are fired out by some radioactive nuclei. The number of electrons needed for a nucleus could be calculated in this way: Consider again the case of carbon 14. We need 14 protons to get the right mass number. But then, as we have seen, the charge is too big. Six of the protons alone supply enough charge to match the value $Z = 6$. The charge of the other 8 protons is unwanted. To cancel these charges, we put in an electron

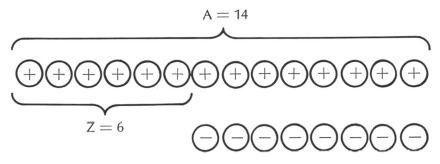

Number of particles = 14 + 8 = 22

Particles needed for carbon 14 under the proton-electron theory of the nucleus

for each of these protons. Moreover, the addition of these electrons does not affect the mass number, since the mass of the electrons is so small. Thus a combination of 14 protons and 8 electrons would have the correct mass number, 14, and the correct atomic number, 6, for carbon 14. In general, if a nucleus has a mass number A, and atomic number Z, a combination of A protons and $A - Z$ electrons would produce the correct mass number and atomic number. Then the total number of particles in the nucleus would be $A + (A - Z) = 2A - Z$.

The proton-electron theory of the nucleus was popular for a while. But then two new discoveries led to its downfall. The

75

discovery of *nuclear spin* uncovered facts which showed that the proton-electron theory was wrong. The discovery of the *neutron* opened the way for a new theory that does fit all the known facts.

NUCLEAR SPIN

We have already seen that an electron and a proton spin like a top, and the spin is equal to $\frac{1}{2}$ (expressed in terms of $h/2\pi$ as a unit). A nucleus also spins like a top, and therefore has angular momentum. The formula for this angular momentum has the form $\sqrt{I(I+1)}$ where I is the largest component the angular momentum can have in any direction. I is known as the *spin* of the nucleus. Because of its spin, the nucleus is like a bar magnet. The magnetism of the nucleus interacts with the magnetism of the spinning and revolving electrons that surround it. This results in a further splitting of the lines
✱ in the spectrum of the atom that contains the nucleus. This splitting of the lines is known as the *hyperfine structure* of the spectrum. The spins of many nuclei have been measured by means of the hyperfine structure of the spectrum. It is found, for example, that the nuclei $_1H^2$, $_3Li^6$, and $_7N^{14}$ have spin 1, and $_{48}Cd^{111}$ has spin $\frac{1}{2}$. $_4Be^9$ has spin 3/2, and $_5B^{10}$ has spin 3.

> ✱ **The spectrum of an atom is like a rainbow with gaps in it. It consists of the colors in a thin beam of mixed light sent out by a glowing mass of atoms of one kind.**
> **The colors in a spectrum are separated and arranged side by side in order of increasing wavelength. Different atoms have different colors in their spectra. The spectrum of an atom contains clues to the structure of the atom.**

The spin of a nucleus is the algebraic sum of the spins of the particles in it plus a spin component contributed by the orbital motion of these particles. The latter is always a whole number. *Consequently the spin of a nucleus is a whole number if and only if the total spin of the particles in it is a whole number.*

The spin of an electron or proton is $\frac{1}{2}$. When small particles are together in a nucleus, their spins are either in the same direction or in the opposite direction. The spins combine to form the total spin of the nucleus according to this rule: Spins in the same direction combine by simple addition, while equal spins that are in opposite directions cancel each other. Thus, 3 particles spinning in the same direction produce a total spin of $\frac{1}{2}+\frac{1}{2}+\frac{1}{2}=3/2$. However, if two of the particles spin in one direction, and one spins

Spins in the same direction   Spins in the opposite direction

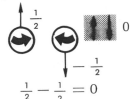

$$\frac{1}{2} + \frac{1}{2} = 1$$ $$\frac{1}{2} - \frac{1}{2} = 0$$

(Shading shows cancellation) Addition of two spins of $\frac{1}{2}$

in the other direction, two of the spins cancel, so the total spin is $\frac{1}{2}$.

The rule for combining spins allows us to predict what kind of spin a nucleus will have if it contains an even number of particles with spin $\frac{1}{2}$. If any of the particles spin in opposite directions, the opposite spins cancel out *in pairs*. So an even number of

All three in the same direction   Spins in the opposite direction

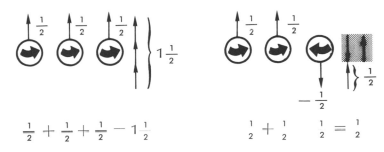

$$\frac{1}{2} + \frac{1}{2} + \frac{1}{2} - 1\frac{1}{2}$$ $$\frac{1}{2} + \frac{1}{2} \quad \frac{1}{2} = \frac{1}{2}$$

(Shading shows cancellation) Addition of three spins of $\frac{1}{2}$

particles have their spins canceled. Subtracting an even number from an even number leaves an even number of particles whose spins are in the same direction. These spins combine by simple addition. So the total spin is an even number times $\frac{1}{2}$. The result is a *whole number. So, if a nucleus contains an even number of particles with spin $\frac{1}{2}$, its total spin is a whole number.*

In the same way we can predict what kind of spin a nucleus will have if it contains an odd number of particles with spin $\frac{1}{2}$. Again, if any of the particles spin in opposite directions, the opposite spins cancel out in pairs. So an even number of particles have their spins canceled. Subtracting an even number from an odd number leaves an odd number of particles whose spins are in the same direction. So the total spin is an odd number times $\frac{1}{2}$. The result is a fractional spin, like $\frac{1}{2}$, or 3/2, or 5/2, and so on. So, *if a*

77

nucleus contains an odd number of particles with spin $\frac{1}{2}$, its total spin is fractional.

We can use this prediction to test the correctness of the proton-electron theory. According to this theory, a nucleus whose mass number is A and whose atomic number is Z contains $2A - Z$ particles. Since $2A$ is an even number, $2A - Z$ is even if Z is even, and it is odd if Z is odd. *So the proton-electron theory requires that a nucleus with even Z have a whole number spin, and that a nucleus with odd Z have a fractional spin. However this requirement does not fit all the facts.* For example, $_1H^2$ has $Z = 1$, but the spin is 1 instead of being fractional. Similarly, the proton-electron theory predicts the wrong kind of spin for $_3Li^6$, and $_7N^{14}$, which also have spin 1, and for $_{48}Cd^{111}$, which has spin $\frac{1}{2}$. *Therefore the proton-electron theory of the nucleus is wrong.*

DISCOVERY OF THE NEUTRON

After the discovery of the nucleus many experiments were performed in which alpha particles were thrown at lightweight atoms. In many cases the alpha particle was absorbed by the nucleus to form a larger nucleus, and then the larger nucleus threw out a proton. For example, when a nitrogen nucleus absorbed an alpha particle, this nuclear reaction took place:

$$_2He^4 + _7N^{14} \rightarrow _9F^{18} \rightarrow _1H^1 + _8O^{17}.$$

($_2He^4$ stands for the alpha particle, and $_1H^1$ stands for the proton.)

When beryllium or boron were the targets at which alpha particles were thrown, the nuclear reaction produced neutral particles instead of protons. At first it was thought that these neutral particles were gamma ray photons. In 1932 Irene Curie and her husband Frederick Joliot showed that these particles can do something that gamma rays never do: they can knock protons out of paraffin. Therefore they could not be gamma rays. The same year Chadwick showed that the particles have the same mass as protons. These new particles are now known as *neutrons*. Since a neutron has no charge and its mass number is 1, we represent it by the symbol $_0n^1$. A neutron spins like a top, and like an electron and proton, it has spin $\frac{1}{2}$.

✱ In 1950 it was shown that when a neutron is not part of a nucleus it is radioactive. A neutron breaks up by beta decay: it fires out an electron, and what is left behind is a proton.

> ✱ A beta particle is an electron. Beta decay means breaking up by firing out an electron.

78

BUILDING BLOCKS OF THE NUCLEUS

Physicists now believe that all nuclei are made of protons and neutrons. Under this proton-neutron theory, a nucleus whose mass number is A is made up of A particles. If the nuclear charge is Z, Z of the A particles are protons and all the rest, that is $A - Z$ of them, are neutrons. For example, the nucleus $_6C^{14}$ consists of 6 protons and 8 neutrons, making a total of 14 particles.

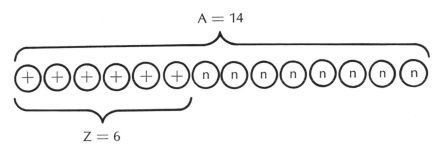

$$A = 14$$

$$Z = 6$$

Number of particles = $6 + 8 = 14$

Particles needed for carbon 14 under the proton-neutron theory of the nucleus

Under this theory the nuclei of all isotopes of the same element have the same number of protons. They differ only in the number of neutrons that they contain. For example, $_1H^1$ has 1 proton and no neutrons, $_1H^2$ has 1 proton and 1 neutron, and $_1H^3$ has 1 proton and 2 neutrons.

In the proton-neutron theory the number of particles in a nucleus is A, the mass number of the nucleus. So this theory predicts that the spin of a nucleus is a whole number if A is even, and it is fractional if A is odd. *This prediction is confirmed by all nuclear spins that have been measured.*

PART 5

The Right Size

The Surfaces and Volumes of Similar Figures

by Galileo Galilei

From a conversation between Salviati and Simplicio

If two things have the same shape but are not the same size, the ratio of their volumes is not the same as the ratio of their surfaces. This fact and some of its consequences are explained in the selections that follow. The first selection is from Dialogues Concerning Two New Sciences *by the great scientist Galileo Galilei. Galileo Galilei was born in Pisa, Italy, on February 15, 1564. He abandoned the study of medicine in favor of mathematics, which he later taught, first at the University of Pisa, and then at the University of Padua. He made many important scientific discoveries, paving the way for the great theories of Isaac Newton. He discovered the law of the pendulum and the law of falling bodies. He also discovered, with a telescope that he made himself, that the moon has a rough surface, with mountains and craters; that Jupiter has moons that revolve around it; that the sun has spots on it and that it rotates; and that Venus goes through phases somewhat similar to the phases of the moon.*

Using his own discoveries in astronomy as evidence, he supported the theory of Copernicus that the earth is a planet that revolves around the sun. This brought him into conflict with the church, which taught that the earth was the center of the universe. In 1633 he was brought before the Inquisition for trial, was forced to renounce the view that the earth revolves around the sun, and was sentenced to prison. He spent the rest of his life as a prisoner in his house. While imprisoned he wrote the Dialogues Concerning Two New Sciences, *which was published in 1638.*

He died on January 8, 1642. Since his death the legend has grown that during his trial, after the threat of a death sentence compelled him to renounce his belief that the earth moves around the sun, he muttered under his breath, "But still it moves."

82

SALV. Now you must know, Simplicio, that it is not possible to diminish the surface of a solid body in the same ratio as the weight, and at the same time maintain similarity of figure. For since it is clear that in the case of a diminishing solid the weight grows less in proportion to the volume, and since the volume always diminishes more rapidly than the surface, when the same shape is maintained, the weight must therefore diminish more rapidly than the surface. But geometry teaches us that, in the case of similar solids, the ratio of two volumes is greater than the ratio of their surfaces; which, for the sake of better understanding, I shall illustrate by a particular case.

Take, for example, a cube two inches on a side so that each face has an area of four square inches and the total area, i. e., the sum of the six faces, amounts to twenty-four square inches; now imagine this cube to be sawed through three times so as to divide it into eight smaller cubes, each one inch on the side, each face one inch square, and the total surface of each cube six square inches instead of twenty-four as in the case of the larger cube. It is evident therefore that the surface of the little cube is only one-fourth that of the larger, namely, the ratio of six to twenty-four; but the volume of the solid cube itself is only one-eighth; the volume, and hence also the weight, diminishes therefore much more rapidly than the surface. If we again divide the little cube into eight others we shall have, for the total surface of one of these, one and one-half square inches, which is one-sixteenth of the surface of the original cube; but its volume is only one-sixty-fourth part. Thus, by two divisions, you see that the volume is diminished four times as much as the surface. And, if the subdivision be continued until the original solid be reduced to a fine powder, we shall find that the weight of one of these smallest particles has diminished hundreds and hundreds of times as much as its surface. And this which I have illustrated in the case of cubes holds also in the case of all similar solids, where the volumes stand in sesquilateral ratio to their surfaces. ✳

✳ Sesquilateral means $1\frac{1}{2}$ times as great.
Galileo is referring to the fact that in the formulas
$V = s^3$ and $A = 6s^2$
for the volume and surface respectively of a cube, the
exponent 3 is $1\frac{1}{2}$ times the exponent 2.

COMPARISON OF THREE CUBES

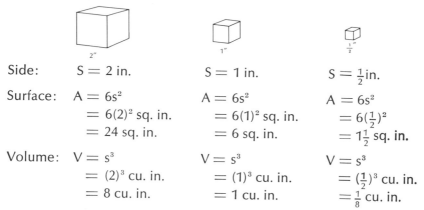

Side:	$S = 2$ in.	$S = 1$ in.	$S = \frac{1}{2}$ in.
Surface:	$A = 6s^2$	$A = 6s^2$	$A = 6s^2$
	$= 6(2)^2$ sq. in.	$= 6(1)^2$ sq. in.	$= 6(\frac{1}{2})^2$
	$= 24$ sq. in.	$= 6$ sq. in.	$= 1\frac{1}{2}$ sq. in.
Volume:	$V = s^3$	$V = s^3$	$V = s^3$
	$= (2)^3$ cu. in.	$= (1)^3$ cu. in.	$= (\frac{1}{2})^3$ cu. in.
	$= 8$ cu. in.	$= 1$ cu. in.	$= \frac{1}{8}$ cu. in.

When the ratio of the sides is 2 to 1,
the ratio of the surfaces is 4 to 1,
and the ratio of the volumes is 8 to 1.

[It is important to see how the surface per cubic inch of volume is changed when a body is made smaller while keeping the same shape. Following Galileo's example, we simplify the problem by assuming that the shape is a cube.

If the side of a cube is 1 inch, its volume is 1 cubic inch. Each face of the cube is 1 square inch, and since there are six faces, the total surface of the cube is 6 square inches. So the surface per cubic inch is 6 square inches.

Now let us consider a cube whose side is 0.1 inch. Its volume is (0.1) (0.1) (0.1) cu. in. $= 0.001$ cu. in. Its surface is 6(0.1)(0.1) sq. in. $= 0.06$ sq. in. This is the surface for a volume of one thousandth of a cubic inch, so to find the surface per cubic inch we must multiply by one thousand. Then, for this cube, the surface per cubic inch is 1000(0.06) sq. in., or 60 sq. in., which is ten times as great as the surface per cubic inch of the one inch cube. Notice that dividing the side of the cube by ten resulted in *multiplying* the surface per cubic inch by ten. If we divide the side of the cube by ten once more, to produce a cube whose side is 0.01 inch, then we again multiply the surface per cubic inch by ten, getting a surface per cubic inch of 600 sq. in. By repeating this procedure over and over again, we get the following further results:

If the side of a cube is 0.001 in., its surface per cubic inch of volume is 6000 sq. in.;

If the side of a cube is 0.0001 in., its surface per cubic inch of volume is 60,000 sq. in.; etc.]

Why Dust Floats

by Irving Adler

Small particles have more surface per unit volume than large objects. This fact plays a part in the answer given below to the question, "Why does dust float?" The answer is taken from the book, Dust *by Irving Adler.*

If you hold a pebble above the ground and then let it go, its weight makes it fall to the ground. In fact, anything that is heavier than air is pulled toward the ground by its own weight. Most dust particles are heavier than air, so they too are pulled down by their weight. In spite of this fact, they are able to float in the air. This is unusual behavior for objects that are heavier than air, and it requires an explanation. What makes it possible for dust particles to float?

We get clues to the answer to this question from some well-known facts of everyday experience. One of these facts is that *a wind exerts a force.* What the weather man calls a light wind is strong enough to make leaves rustle or move a weather vane. A fresh wind will make small trees sway. A strong wind pushes you so hard it becomes difficult to walk against it. A whole gale is strong enough to knock down trees.

A second fact is that *moving bodies make their own wind.* If you ride on a bicycle on a windless day, as soon as you start moving you feel a wind in your face. This wind is the result of the fact that, as you ride the bicycle, you move through the air. While the air stands still and you move through it, the air flows past you just as it does when you stand still and the air moves in an ordinary wind.

A third important fact can be observed by means of a simple experiment. Take two sheets of paper of the same size and

85

crumple up one of them to form a tight little ball. Hold the open sheet in one hand, and the crumpled sheet in the other. Then drop them to the ground from the same height at the same time. The crumpled sheet will fall quickly, but the open sheet will float down slowly, swaying from side to side. Both sheets create their own wind as they fall, but the wind holds the open sheet back more, because the open sheet has more surface for the wind to blow against. *The bigger the surface against which a wind blows, the stronger is the force with which the wind pushes it.* A flier takes advantage of this fact when he uses a parachute. As he falls through the air after jumping from a plane, he makes his own wind. If he is not wearing a parachute, the wind slows him down very little, and he hurtles to the ground, falling faster and faster to a certain death. But if he wears a parachute, the open parachute gives the wind a larger surface to blow against. The greater force of the wind resistance holds back his fall, and he floats gently to the ground. In fact, if he happens to fall into an updraft of air, the upward-flowing air, pushing against the parachute, could even lift him up, in spite of his weight.

A fourth important fact can be noticed any time you break something. *When a solid object is broken, new surfaces are formed along the breaks.* So, when an object is broken into pieces, the total surface of the object is increased. The smaller the pieces

New surface ⌐

 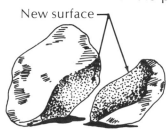

into which it is broken, the more surface it gets. Because of this fact, when a small piece of matter is ground into dust, although its weight remains the same, its surface is increased tremendously. For example, if a pebble shaped like a cube is one centimeter wide (almost a half inch), its surface area is about one square inch. But

✱ if it is ground up into a thousand billion billion small pieces, the total surface of all these particles of dust would be about an *acre*.

✱ 10^{20}

These four facts serve to explain why particles of dust can float. If a pebble is dropped in the air, it makes its own wind as it falls. The wind, pushing against the surface of the pebble, tends to hold it back. But, since the surface of the pebble is small, the wind doesn't push it very hard, and the pebble falls quickly to the ground. But if the pebble is ground into dust before it is dropped, it offers the wind more surface to blow against. The wind, blowing against the greater surface, pushes the particles harder than it would if they were all in one lump. If the particles are small enough, the wind resistance is great enough to make them settle slowly. In an updraft, the force of the wind may even be strong enough to push the particles up in spite of their weight. *Because grinding a solid gives it more surface, grinding it into dust is like giving it a parachute.* That is why particles of dust can float in the air.

Gentle currents of air are strong enough to raise dust particles off the ground and blow them about. When the currents die down, the heavier particles begin to settle. The wind the particles make as they fall makes them settle slowly, but they do settle, as every housewife knows to her sorrow. But some of the particles of dust are so small that they never settle at all. In addition to the wind resistance which tends to hold them up, they are pushed about by collisions with molecules of air. The upward pushes they get from the molecules that they strike are strong enough to prevent them from settling. The average width of these particles that hang in the air forever is about one half a micron. Their presence ✷ in the air is responsible for many of the interesting things that happen in the air.

✷ **A micron is one millionth of a meter. A single inch contains about 25,400 microns.**

87

Surface and Shape

The preceding selections stressed the fact that you can increase the surface per unit volume that a thing has by making it smaller without changing its shape. There is also another way of increasing the surface per unit volume, by *changing the shape without changing the volume.*

Consider, for example, a cube whose side is one inch long. As we have seen before, its volume is one cubic inch, and its surface is 6 square inches. If we cut the cube into four pieces, as shown in the diagram, and pile the pieces one on top of the other, we get a prism that is $\frac{1}{2}$ inch wide on each side and is 4 inches high. The volume of this figure, which is taller and thinner than the cube is the same one cubic inch we had before. But its surface is *more* than 6 square inches. In fact, it has two faces, the upper and lower base, that contain $\frac{1}{4}$ square inch each; and four faces, the side faces, that contain 2 square inches each. So the total surface is $8\frac{1}{2}$ square inches. In general, *the taller and thinner a thing is, the more surface it has per unit volume.*

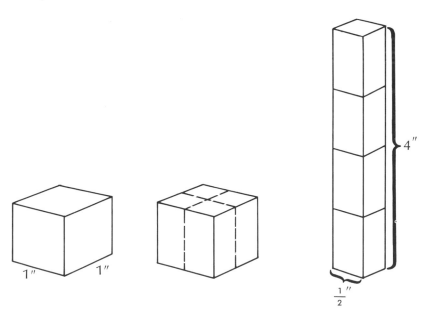

1″   1″   4″ $\frac{1}{2}$″

On Being the Right Size

by J. B. S. Haldane

In the next selection, a famous biologist tells about many interesting properties of living things that depend on how surface per unit volume varies with size and shape.

The most obvious differences between different animals are differences of size, but for some reason the zoologists have paid singularly little attention to them. In a large textbook of zoology before me I find no indication that the eagle is larger than the sparrow, or the hippopotamus bigger than the hare, though some grudging admissions are made in the case of the mouse and the whale. But yet it is easy to show that a hare could not be as large as a hippopotamus, or a whale as small as a herring. For every type of animal there is a most convenient size, and a large change in size inevitably carries with it a change of form.

Let us take the most obvious of possible cases, and consider a giant man sixty feet high — about the height of Giant Pope and Giant Pagan in the illustrated *Pilgrim's Progress* of my childhood. These monsters were not only ten times as high as Christian, but ten times as wide and ten times as thick, so that their total weight was a thousand times his, or about eighty to ninety tons. Unfortunately the cross sections of their bones were only a hundred times those of Christian, so that every square inch of giant bone had to support ten times the weight borne by a square inch of human bone. As the human thigh-bone breaks under about ten times the human weight, Pope and Pagan would have broken their thighs every time they took a step. This was doubtless why they were sitting down in the picture I remember. But it lessens one's respect for Christian and Jack the Giant Killer.

89

To turn to zoology, suppose that a gazelle, a graceful little creature with long thin legs, is to become large, it will break its bones unless it does one of two things. It may make its legs short and thick, like the rhinoceros, so that every pound of weight has still about the same area of bone to support it. Or it can compress its body and stretch out its legs obliquely to gain stability, like the giraffe. I mention these two beasts because they happen to belong to the same order as the gazelle, and both are quite successful mechanically, being remarkably fast runners.

Gravity, a mere nuisance to Christian, was a terror to Pope, Pagan, and Despair. To the mouse and any smaller animal it presents practically no dangers. *You can drop a mouse down a thousand-yard mine shaft; and, on arriving at the bottom, it gets a slight shock and walks away,* provided that the ground is fairly soft. A rat is killed, a man is broken, a horse splashes. For the resistance presented to movement by the air is proportional to the surface of the moving object. Divide an animal's length, breadth, and height each by ten; its weight is reduced to a thousandth, but its surface only to a hundredth. So the resistance to falling in the case of the small animal is relatively ten times greater than the driving force.

An insect, therefore, is not afraid of gravity; it can fall without danger, and can cling to the ceiling with remarkably little trouble. It can go in for elegant and fantastic forms of support like that of the daddy-long-legs. But there is a force which is as formidable to an insect as gravitation to a mammal. This is surface tension. A man coming out of a bath carries with him a film of water of about one-fiftieth of an inch in thickness. This weighs roughly a pound. A wet mouse has to carry about its own weight of water. A wet fly has to lift many times its own weight and, as every one knows, a fly once wetted by water or any other liquid is in a very serious position indeed. **An insect going for a drink is in as great danger as a man leaning out over a precipice in search of food. If it once falls into the grip of the surface tension of the water — that is to say, gets wet — it is likely to remain so until it drowns.** A few insects, such as waterbeetles, contrive to be unwettable; the majority keep well away from their drink by means of a long proboscis.

Of course tall land animals have other difficulties. They have to pump their blood to greater heights than a man and, therefore, require a larger blood pressure and tougher blood-vessels. A great many men die from burst arteries, especially in the brain, and this danger is presumably still greater for an elephant or a giraffe.

But animals of all kinds find difficulties in size for the following reason. A typical small animal, say a microscopic worm or rotifer, has a smooth skin through which all the oxygen it requires can soak in, a straight gut with sufficient surface to absorb its food, and a simple kidney. Increase its dimensions tenfold in every direction, and its weight is increased a thousand times, so that if it is to use its muscles as efficiently as its miniature counterpart, it will need a thousand times as much food and oxygen per day and will excrete a thousand times as much of waste products.

Now if its shape is unaltered its surface will be increased only a hundredfold, and ten times as much oxygen must enter per minute through each square millimetre of skin, ten times as much food through each square millimetre of intestine. When a limit is reached to their absorptive powers their surface has to be increased by some special device. For example, a part of the skin may be drawn out into tufts to make gills or pushed in to make lungs, thus increasing the oxygen-absorbing surface in proportion to the animal's bulk. A man, for example, has a hundred square yards of lung. Similarly, the gut, instead of being smooth and straight, becomes coiled and develops a velvety surface, and other organs increase in complication. *The higher animals are not larger than the lower because they are more complicated. They are more complicated because they are larger.* Just the same is true of plants. The simplest plants, such as the green algae growing in stagnant water or on the bark of trees, are mere round cells. The higher plants increase their surface by putting out leaves and roots. *Comparative anatomy is largely the story of the struggle to increase surface in proportion to volume.*

Some of the methods of increasing the surface are useful up to a point, but not capable of a very wide adaptation. For example, while vertebrates carry the oxygen from the gills or lungs all over the body in the blood, insects take air directly to every part of their body by tiny blind tubes called tracheae which open to the surface at many different points. Now, although by their breathing movements they can renew the air in the outer part of the tracheal system, the oxygen has to penetrate the finer branches by means of diffusion. Gases can diffuse easily through very small distances, not many times larger than the average length travelled by a gas molecule between collisions with other molecules. But when such vast journeys — from the point of view of a molecule — as a quarter of an inch have to be made, the process becomes slow. **So the portions**

91

of an insect's body more than a quarter of an inch from the air would always be short of oxygen. **In consequence hardly any insects are much more than half an inch thick.** Land crabs are built on the same general plan as insects, but are much clumsier. Yet like ourselves they carry oxygen around in their blood, and are therefore able to grow far larger than any insects. If the insects had hit on a plan for driving air through their tissues instead of letting it soak in, they might well have become as large as lobsters, though other considerations would have prevented them from becoming as large as man.

Exactly the same difficulties attach to flying. It is an elementary principle of aeronautics that the minimum speed needed to keep an aeroplane of a given shape in the air varies as the square root of its length. If its linear dimensions are increased four times, it must fly twice as fast. Now the power needed for the minimum speed increases more rapidly than the weight of the machine. So the larger aeroplane, which weighs sixty-four times as much as the smaller, needs one hundred and twenty-eight times its horsepower to keep up. Applying the same principles to the birds, we find that the limit to their size is soon reached. An angel whose muscles developed no more power weight for weight than those of an eagle or a pigeon would require a breast projecting for about four feet to house the muscles engaged in working its wings, while to economize in weight, its legs would have to be reduced to mere stilts. Actually a large bird such as an eagle or kite does not keep in the air mainly by moving its wings. It is generally to be seen soaring, that is to say balanced on a rising column of air. And even soaring becomes more and more difficult with increasing size. Were this not the case eagles might be as large as tigers and as formidable to man as hostile aeroplanes.

But it is time that we passed to some of the advantages of size. One of the most obvious is that it enables one to keep warm. All warm-blooded animals at rest lose the same amount of heat from a unit area of skin, for which purpose they need a food-supply proportional to their surface and not to their weight. *Five thousand mice weigh as much as a man. Their combined surface and food or oxygen consumption are about seventeen times a man's. In fact a mouse eats about one quarter its own weight of food every day, which is mainly used in keeping it warm.* For the same reason small animals cannot live in cold countries. In the arctic regions there are no reptiles or amphibians, and no small mammals. The smallest

92

mammal in Spitzbergen is the fox. The small birds fly away in the winter, while the insects die, though their eggs can survive six months or more of frost. The most successful mammals are bears, seals, and walruses.

Similarly, the eye is a rather inefficient organ until it reaches a large size. The back of the human eye on which an image of the outside world is thrown, and which corresponds to the film of a camera, is composed of a mosaic of 'rods and cones' whose diameter is little more than a length of an average light wave. Each ✳ eye has about half a million, and for two objects to be distinguishable their images must fall on separate rods or cones. It is obvious that with fewer but larger rods and cones we should see less distinctly. If they were twice as broad two points would have to be twice as far apart before we could distinguish them at a given distance. But if their size were diminished and their number increased we should see no better. For it is impossible to form a definite image smaller than a wave-length of light. Hence a mouse's eye is not a small-scale model of a human eye. Its rods and cones are not much smaller than ours, and therefore there are far fewer of them. A mouse could not distinguish one human face from another six feet away. **In order that they should be of any use at all the eyes of small animals have to be much larger in proportion to their bodies than our own. Large animals on the other hand only require relatively small eyes, and those of the whale and elephant are little larger than our own.**

✳ **The length of an average light wave is, of course, tiny — somewhere between .035 and .075 millimeters.**

For rather more recondite reasons the same general principle holds true of the brain. If we compare the brain-weights of a set of very similar animals such as the cat, cheetah, leopard, and tiger, we find that as we quadruple the body-weight the brain-weight is only doubled. The larger animal with proportionately larger bones can economize on brain, eyes, and certain other organs.

Such are a very few of the considerations which show that for every type of animal there is an optimum size. Yet although Galileo demonstrated the contrary more than three hundred years ago, people still believe that if a flea were as large as a man it could jump a thousand feet into the air. *As a matter of fact the height to which an animal can jump is more nearly independent of its size than proportional to it.* A flea can jump about two feet, a man about five. To jump a given height, if we neglect the resistance of the air, requires an expenditure of energy proportional to the jumper's weight. But if the jumping muscles form a constant fraction of the animal's body, the energy developed per ounce of muscle is independent of the size, provided it can be developed quickly enough in the small animal. As a matter of fact an insect's muscles, although they can contract more quickly than our own, appear to be less efficient; as otherwise a flea or grasshopper could rise six feet into the air.

And just as there is a best size for every animal, so the same is true for every human institution. In the Greek type of democracy all the citizens could listen to a series of orators and vote directly on questions of legislation. Hence their philosophers held that a small city was the largest possible democratic state. The English invention of representative government made a democratic nation possible, and the possibility was first realized in the United States, and later elsewhere. With the development of broadcasting it has once more become possible for every citizen to listen to the political views of representative orators, and the future may perhaps see the return of the national state to the Greek form of democracy. Even the referendum has been made possible only by the institution of daily newspapers.

To the biologist the problem of socialism appears largely as a problem of size. The extreme socialists desire to run every nation as a single business concern. I do not suppose that Henry Ford would find much difficulty in running Andorra or Luxembourg on a socialistic basis. He has already more men on his pay-roll than their population. It is conceivable that a syndicate of Fords, if we could find them, would make Belgium Ltd. or Denmark Inc. pay their way. But while nationalization of certain industries is an obvious possibility in the largest of states, I find it no easier to picture a completely socialized British Empire or United States than an elephant turning somersaults or a hippopotamus jumping a hedge.

PART 6

Space Fillers

Space-Filling Figures

by Irving Adler

*There are infinitely many kinds of regular polygons. But only three
kinds can be used like tiles on a floor to fill out a plane. Irving
Adler explains why in the next selection, taken
from* A New Look at Geometry.

The Pythagoreans initiated the theory of space-filling figures. The
main problem of this theory is to find figures that can be repeated,
like tiles on a floor, to fill out a plane. Three simple solutions to the
problem are shown below. In the first one, the figure that is repeated is an equilateral triangle; in the second one, it is a square;
in the third one, it is a regular hexagon. In all three solutions, the
figure that is repeated is a regular polygon. It is interesting that these
are the only solutions to the problem in which the figure that is
repeated is a regular polygon. To be able to prove this assertion, let
us first review quickly some elementary facts about regular polygons.

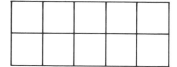

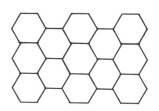

A regular polygon is one that has equal sides and equal angles. To draw a regular polygon, it suffices to divide a circle into three or more equal parts and join the successive points of division. Therefore a regular polygon can be drawn with n sides for any integral value of n greater than or equal to 3. Let us call a regular polygon with n sides a regular n-gon. It is obvious that a regular n-gon has n angles.

It is easy to calculate the number of degrees in each angle of a regular n-gon by first calculating the number of degrees in each exterior angle formed by extending one side. Let us denote by x the number of degrees in the exterior angle. To calculate x we take one exterior angle at each vertex of the n-gon, as shown in the diagram below, and then add them up. To add the angles, we use a hand of a clock in this way: Start with the hand placed parallel to the horizontal side of angle 1. Rotate the hand counterclockwise until it has swept out an angle equal to angle 1. The hand will end up parallel to the other side of angle 1, whose extension is a side of angle 2.

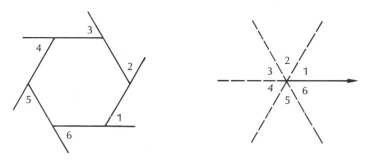

Now rotate the hand until it sweeps out an angle equal to angle 2. In its new position, the hand is parallel to a side of angle 3. Continue in this way, sweeping out with the hand an angle equal to each of the exterior angles of the n-gon in succession. In its final position the hand will be back where it started from. That is, n successive rotations of x degrees each add up to one complete rotation, or 360 degrees. Then, since $nx = 360$, $x = \dfrac{360}{n}$. At each vertex of the n-gon, an angle of the n-gon and the exterior angle next to it add up to 180 degrees. Consequently the number of degrees in each angle of a regular n-gon is $180 - \dfrac{360}{n}$, or $180\left(1 - \dfrac{2}{n}\right)$.

Now we are ready to consider the problem of finding all the values of n for which repetitions of a regular n-gon can fill out a plane. Let p be the number of n-gons that occur at each vertex. Then there will be p angles at each vertex, and since they fill out the space in the plane around the vertex, their sum is 360 degrees. Consequently the number of degrees in each of them is $\dfrac{360}{p}$. But we know already that the number of degrees in each angle of a regular n-gon is $180\left(1-\dfrac{2}{n}\right)$. Equating these two expressions, and dividing by 180, we get the equation;

(1)
$$1-\frac{2}{n}=\frac{2}{p}\text{, or}$$

(2)
$$1-\frac{2}{n}-\frac{2}{p}=0.$$

If we multiply equation (2) by np, we get

(3)
$$np-2p-2n=0.$$

If we add 4 to both sides of equation (3) we get

(4)
$$np-2p-2n+4=4.$$

Factoring the left-hand side of equation (4) we get

(5)
$$(n-2)(p-2)=4.$$

Since $n-2$ and $p-2$ are whole numbers, and their product is 4, we get all possible values of n and p by equating the pair $n-2$, $p-2$ to all possible pairs of whole numbers whose product is 4. These possible pairs are 4, 1; 2, 2; and 1, 4. Consequently there are only three solutions, given by the three pairs of equations:

$$\begin{cases}n-2=4\\p-2=1\end{cases}\qquad\begin{cases}n-2=2\\p-2=2\end{cases}\qquad\begin{cases}n-2=1\\p-2=4\end{cases}$$

The first pair yields the solution $n=6$, $p=3$. The second pair yields the solution $n=4$, $p=4$. The third pair yields the solution $n=3$, $p=6$. Therefore there are only three ways of filling a plane by repeating a regular n-gon: use equal regular 6-gons (regular hexagons) with three at each vertex, or equal regular 4-gons (squares) with four at each vertex, or equal regular 3-gons (equilateral triangles) with six at each vertex.

On the Sagacity of Bees in Building Their Cells

by Pappus

*Bees use the regular hexagon as a space-filling figure when they
build their honeycombs. Pappus, a Greek mathematician who lived
in Alexandria about 300 A.D., comments on this fact in the
following selection, taken from his book,* Mathematical
Collection. *The translation is by T.L. Heath, in
his* History of Greek Mathematics.

It is of course to men that God has given the best and most perfect
notion of wisdom in general and of mathematical science in particular, but a partial share in these things he allotted to some of the
unreasoning animals as well. To men, as being endowed with reason, he vouchsafed that they should do everything in the light of
reason and demonstration, but to the other animals, while denying
them reason, he granted that each of them should, by virtue of a
certain natural instinct, obtain just so much as is needful to support
life. This instinct may be observed to exist in very many other
species of living creatures, but most of all in bees. In the first place
their orderliness and their submission to the queens who rule in
their state are truly admirable, but much more admirable still is their
emulation, the cleanliness they observe in the gathering of honey,
and the forethought and housewifely care they devote to its custody.
Presumably because they know themselves to be entrusted with the
task of bringing from the gods to the accomplished portion of man-
kind a share of ambrosia in this form, they do not think it proper to
pour it carelessly on ground or wood or any other ugly and irregular
material; but, first collecting the sweets of the most beautiful flowers which grow on the earth, they make from them, for the recep-
tion of the honey, the vessels which we call honeycombs, (with
cells) all equal, similar and contiguous to one another, and hexago-

nal in form. And that they have contrived this by virtue of a certain geometrical forethought we may infer in this way. They would necessarily think that the figures must be such as to be contiguous to one another, that is to say, to have their sides common, in order that no foreign matter could enter the interstices between them and so defile the purity of their produce. Now only three rectilineal figures would satisfy the condition, I mean regular figures which are equilateral and equiangular; for the bees would have none of the figures which are not uniform. ... There being then three figures capable by themselves of exactly filling up the space about the same point, the bees by reason of their instinctive wisdom chose for the construction of the honeycomb the figure which has the most angles, because they conceived that it would contain more honey than either of the two others.

Bees, then, know just this fact which is of service to themselves, that the hexagon is greater than the square and the triangle and will hold more honey for the same expenditure of material used in constructing the different figures.

There are other space-filling patterns that are not regular polygons. Although there is no limit to how many of these a creative artist can invent, mathematical analysis shows that the basic types of these patterns are limited in number. There are exactly seventeen of them. It is interesting that the Alhambra, built by the Moors in Spain long before this mathematical analysis had been made, includes all seventeen types of space-filling patterns among its wall decorations. Simple examples of four types of space-filling patterns are shown, reproduced from Regular Figures *by L. Fejes Toth.*

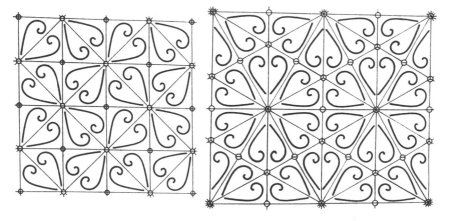

The space-filling patterns shown below come from different continents. The first one is a window of a mosque built in Cairo, Egypt, during the fourteenth century. The other two are Chinese windows. (Reproduced from Symmetry, by Hermann Weyl.)

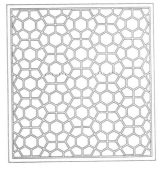

Space-Filling Drawings Borrowed From Nature

by M. C. Escher

*The Moors used geometric figures to make the space-filling
patterns that decorate the Alhambra. The Dutch artist,
M. C. Escher, has added a new twist to space-filling by making
drawings of animals, people, angels, devils, and other interesting
things which he fits together to fill a plane. His drawings, while
fascinating in themselves, have a special interest to crystallographers,
the scientists who study the structure of crystals, because crystals
are examples of three-dimensional space-filling arrangements
of molecules. A collection of Mr. Escher's drawings was published
for the International Union of Crystallography in the book entitled
Symmetry Aspects of M. C. Escher's Periodic Drawings. We
reproduce from this book Mr. Escher's preface and
two of his drawings.*

Many of the brightly coloured, tile-covered walls and floors of the
Alhambra in Spain show us that the Moors were masters in the art
of filling a plane with similar interlocking figures, bordering each
other without gaps.

**What a pity that their religion forbade them to make
images! It seems to me that they sometimes have been very near to
the development of their elements into more significant figures than
the abstract geometric shapes which they moulded. But no Moorish
artist has, as far as I know, ever dared (or did he not hit on the idea?)
to use as building components: concrete, recognizable figures, borrowed from nature, such as fishes, birds, reptiles or human beings.
This is hardly believable because the recognizability of my own
plane-filling elements not only makes them more fascinating, but
this property is the very reason of my long and still continuing activ-
ity as a designer of periodic drawings.**

102

Another important question is shade contrast. For the Moors it was natural to compose their tiled surfaces with mutually contrasting, different-coloured pieces of majolica. Likewise I myself have always used contrasting shades as a simple necessity, as a logical means of visualising the adjacent components of my patterns.

These two main rules could briefly be formulated as follows: without recognizability no meaning and without shade contrast no visibility.

I often wondered at my own mania of making periodic drawings. Once I asked a friend of mine, a psychologist, about the reason of my being so fascinated by them, but his answer: that I must be driven by a primitive, prototypical instinct, does not explain anything.

What can be the reason of my being alone in this field? Why do none of my fellow-artists seem to be fascinated as I am by these interlocking shapes? Yet their rules are purely objective ones, which every artist could apply in his own personal way!

My first periodic woodcut was made in 1922. The original woodblock presents a collection of eight different human heads which can be printed and multiplied by translation.

In the course of the years I designed about a hundred and fifty of these tessellations. In the beginning I puzzled quite ✳ *instinctively, driven by an irresistible pleasure in repeating the same forms, without gaps, on a piece of paper. These first drawings were tremendously time-devouring because I had never heard of crystal-* ✳ *lography; so I did not even know that my game was based on rules which have been scientifically investigated. Nor had I visited the Alhambra at that time.*

✳ **Tessellations are space-filling designs, like mosaics, where shapes are ordered into a coherent pattern, and fill a plane.**

✳ **Crystallography is the science governing the forms, structures and shapes which make up crystals or compose tessellations.**

Many years later, in 1935, I came for the first time in contact with crystallographic theories, which I seriously tried to understand. But they were mostly too difficult for my untrained mind and on the other hand they took no account of the shade contrasts which for me are indispensable. So in 1942 I came to formulate a

103

personal layman's theory on colour symmetries which I illustrated with many explanatory figures.

Though the text of scientific publications is mostly beyond my means of comprehension, the figures with which they are illustrated bring me occasionally on the track of new possibilities for my work. It was in this way that a fruitful contact could be established between mathematicians and myself.

The dynamic action of making a symmetric tessellation is done more or less unconsciously. While drawing I sometimes feel as if I were a spiritualist medium, controlled by the creatures which I am conjuring up. It is as if they themselves decide on the shape in which they choose to appear. They take little account of my critical opinion during their birth and I cannot exert much influence on the measure of their development. They usually are very difficult and obstinate creatures.

The border line between two adjacent shapes having a double function, the act of tracing such a line is a complicated business. On either side of it, simultaneously, a recognizability takes shape. But the human eye and mind cannot be busy with two things at the same moment and so there must be a quick and continuous jumping from one side to the other. But this difficulty is perhaps the very moving-spring of my perseverance.

This publication of my periodic drawings represents for me a crown on an important part of my life's work. I am deeply indebted and grateful to Professor G. H. MacGillavry, because the book would never have been accomplished without her kind and flattering interest in my regular plane-filling mania.

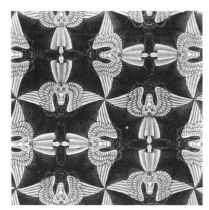

PART 7

Squares

Square and Circle
by Lancelot Hogben

Mathematics and civilization grew up together. The way in which they helped each other grow is described by Lancelot Hogben in his book, The Wonderful World of Mathematics. *The next selection, Hogben's chapter on Babylonian mathematics, includes a method of using squares to multiply numbers, and a method of using squares to get an approximate value of pi.*

Lancelot Hogben is a British biologist whose interests and writings go far beyond his specialty. He has written many popular books on science, mathematics, and language. He was the first Vice Chancellor of the University of Guyana.

A thousand miles east of the delta of the river Nile are two other great rivers, the Tigris and Euphrates. Between and beside their banks, in the land called Mesopotamia, there grew up another civilization, at least as ancient as that of Egypt.

Historians refer to this civilization at different stages of its development as Sumerian, Chaldean, Assyrian, and Babylonian. It was in some ways very similar to the Egyptian. In both, the priestly sky watchers and calendar keepers were the ruling class and both made astonishing progress in astronomy. By about 2000 B.C. the priests in these lands had built up temple libraries where they recorded their knowledge in a secret script that ordinary men could not read.

There the resemblance between the two civilizations ends.

Mesopotamia, unlike early Egypt, had a considerable foreign trade. It had no wood of its own suitable for building, no silk of its own to clothe kings and princes, no spices for the dishes of the wealthy, few precious metals from which to make vessels for the temples. To meet all these needs, merchants with caravans of asses or camels traveled through mountain pass and over desert, going westward to Lebanon for cedar wood, northward into Asia Minor for gold, silver, lead, or copper, and eastward possibly as far as India and China for silks, dyes, spices, and jewels.

A merchant, selling the produce of his fields, may be content to measure his wares roughly and to sell them by the donkey load; but a merchant dealing in more costly goods needs to be far more precise.

Thus, in Mesopotamia, scales and standard weights came into common use, and the merchant weighed his heavy goods in talents (roughly 55 pounds) and his precious wares in shekels (rather less than one third of an ounce). But the merchant also needed to find something that anyone and everyone would accept as payment for goods. There was one thing that almost everybody would accept: barley. For many years, barley was the workman's wage.

Assyrian and Babylonian weights.

What was left over after he had made his bread and brewed his beer he could exchange for other things. So the early Mesopotamian merchants, when they set off to trade with other lands, loaded their asses and camels with barley to pay for the goods they intended to buy.

As time went on they discovered that silver, much lighter and easier to handle than barley, would also be acceptable almost everywhere. At first, they would carry small quantities of it and weigh it out as necessary. Later, they sidestepped the constant trouble of weighing by casting small bars of silver each stamped to

show its weight. Although not much like our modern coins in appearance, they were the world's first money.

Here, for the first time, was a kind of wealth that a man could save without fear that it would go bad. He could also lend it out and charge interest on it, as the usurers of the Bible story did. To do this, as when buying or selling, he would need to keep accounts.

In this task, the merchants of Mesopotamia were handicapped by a clumsy script and bulky writing material. They wrote with pointed sticks on tablets of soft clay; and they had to bake the tablets hard in the sun to hold the impression. The process must have slowed down writing considerably, but it made the finished tablets difficult to destroy. In recent years archaeologists have found thousands of them with wedge-shaped signs, called cuneiform, still clearly written on them.

It needed masterly detective work to decipher the writing, partly because the signs, at first sight, all look very much alike, partly because different scribes used signs in different ways.

What stood for 10 in one place might stand for 60 in another; what stood for 100 (10 × 10) on one tablet might mean 3600 (60 × 60) on the next.

Although Mesopotamia had elaborate systems of weights and measures, and although it was the first home of money, its methods of keeping written accounts remained at a very crude level. Fortunately for them, however, the merchants had a way of calculating without written numbers.

Like the Egyptians, they set out pebbles in grooves in the sand, each pebble in the first groove standing for 1, each in the second groove for 10, each in the third for 100, and so on. The diagrams show how the merchants used this device, called the *abacus,* for adding up their accounts, long before there were any rules for written arithmetic.

Notice how the number value of a pebble grows as you move it from groove to groove: 1 in the first groove, 10 × 1 in the second, 10 × 10 × 1 in the third, and so on. When the value of a pebble in each groove or column is 10 times greater than the one before it, we now say that 10 is the base.

Most ancient number systems used a base of 10, probably because most people first learned to count on the fingers of their two hands. But there is nothing magic about the number 10. It is just as easy to work with a quite different base.

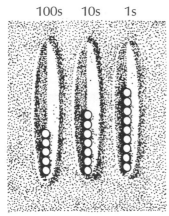

A. To add 579 to 152, first set out 579: 5 hundreds, 7 tens, 9 ones.

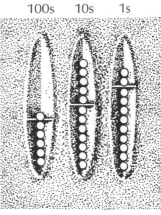

B. Add pebbles to show 152: 1 hundred, 5 tens, 2 ones. Ones column has 11.

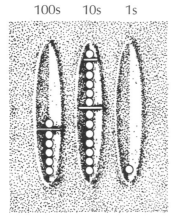

C. Carry 1 from ones column to tens; throw 9 out. Tens column has 13.

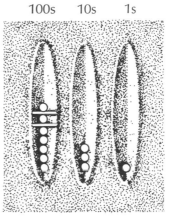

D. Carry 1 from tens to hundreds; throw 9 out. Pebbles give answer: 731.

The Egyptians used 10 as a base, and consistently used separate signs for 1, 10, 100, 1000, and so on, repeated to recall the number of pebbles in the corresponding groove. Although the people of Mesopotamia also used a base of 10, they sometimes used a base of 60. A trace of this has come down to us. In measuring time, we still divide the hour into 60 minutes and the minute into 60 seconds. Navigators, in measuring distance, still divide each degree of longitude or latitude into 60 minutes, and each minute into 60 seconds. The Mayas of Central America, who may have used both toes and fingers for counting, used a base of 20 except in time reckoning.

109

Mesopotamia was not the only country handicapped by clumsy signs for numbers. The same thing was true of most civilized lands until a few centuries ago, and so the habit of using the abacus spread in the course of time over most of the world.

The abacus used in ancient Rome was a metal plate with two sets of parallel grooves, one below the other. The lower set held four pebbles in each groove and the upper set only one. The pebble in an upper groove was worth five times as much as a pebble in the corresponding lower groove. Thus the operator could show any number up to nine in each complete column. At the right of the metal plate there was a separate set of grooves used for working with fractions. The word the Romans used for a pebble was *calculus*, from which we get our own word calculate.

The abacus was not the only shortcut to calculation known to the merchants and traders of Mesopotamia. Among the thousands of clay tablets that archaeologists have unearthed from a temple library near the banks of the Euphrates, some are tables of multiplication and addition, and others are tables of the squares of numbers. The square of a number simply means a number multiplied by itself, such as $2 \times 2 = 4$, which we now write as 2^2, or $5 \times 5 = 25$, which we now write as 5^2.

It seems likely that the priests of those days had discovered a way of using square tables that enabled them to multiply any two numbers together without using the abacus. Here, for example, is how they would multiply 102 by 96.

STEP 1 Add 102 to 96 and divide the result by 2 to find the average 99

STEP 2 Take 96 from 102; divide the result by 2, to find half the difference between the two numbers 3

STEP 3 Look up in the table the square of 99 and you at once see it is 9801

STEP 4 Look up in the table the square of 3 and you at once see it is 9

STEP 5 Take 9 from 9801 and you find the correct answer ..9792

A square shows the results of multiplying numbers by themselves.

	1	2	3	4	5	6	7
1	1	2	3	4	5	6	7
2	2	4	6	8	10	12	14
3	3	6	9	12	15	18	21
4	4	8	12	16	20	24	28
5	5	10	15	20	25	30	35
6	6	12	18	24	30	36	42
7	7	14	21	28	35	42	49

If we understand this method, we can multiply any two numbers together in the same way. When we multiply one number by another, the result is always equal to the square of their average minus the square of half the difference between them. Yet square-table multiplication was never as widespread as the use of the abacus. Long after the time of Columbus, some merchants and shopkeepers of western Europe still used counting boards worked on much the same principle as the abacus. In some parts of Asia the modern businessman still uses the abacus today and works with great speed.

[Here is the reasoning behind the square-table method of multiplication:

Let x and y be the numbers to be multiplied.

Their average is $\dfrac{x + y}{2}$.

The square of their average is $\dfrac{x^2 + 2xy + y^2}{4}$.

Half the difference between them is $\dfrac{x - y}{2}$.

The square of the latter is $\dfrac{x^2 - 2xy + y^2}{4}$.

Finally,

$$\dfrac{x^2 + 2xy + y^2}{4} - \dfrac{x^2 - 2xy + y^2}{4}$$

$$= \dfrac{(x^2 + 2xy + y^2) - (x^2 - 2xy + y^2)}{4}$$

$$= \dfrac{x^2 + 2xy + y^2 - x^2 + 2xy - y^2}{4}$$

$$= \dfrac{4xy}{4} = xy.]$$

About 6000 years ago some unknown citizen of ancient Mesopotamia made one of the greatest inventions of all time: the wheel. At first it was no more than a solid disk of wood with a hole in the middle to allow it to revolve around a fixed axle. By the time the Babylonians and Assyrians built their trading carts and war chariots it had become much more like the wheel of the farm cart still sometimes seen today, with rim, spokes, and hub.

Mesopotamia found other uses for the wheel, too. By placing his clay on a turning wheel, the potter could mold his vessels more accurately. With the help of pulley wheels, builders and engineers could raise heavy weights more easily.

It is tempting to suppose that the Mesopotamians, with their knowledge of wheels, learned a good deal about the geometry of the circle; but in fact they made no more progress than the Egyptians — probably not as much. The Egyptians estimated that the boundary, or circumference, of a circle is 3.14 times as long as its diameter. Since this ratio of circumference to diameter (or of area to square of radius), now called by the Greek letter π (pronounced pi), is approximately 3.1416, this was quite a close estimate. The Mesopotamians were commonly content to use the more convenient but less precise value 3.0.

How these early peoples arrived at any value for π, however crude, is not certain; but some of their inscriptions give us a clue. By drawing the smallest square that can enclose a circle and the largest that can fit inside it, they could see that the boundary of a circle lies between the boundaries of the two squares, and it happens that the average boundary of the two squares is almost 3.4 times the diameter of the circle. By drawing one hexagon outside, and another inside the circle, they could have obtained an even closer estimate.

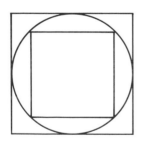
Size of circle is roughly half the sum of the outer and inner figure.
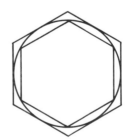

If the priests of Mesopotamia knew less about the circle than did those of Egypt, their knowledge of practical geometry was not inferior. In astronomy, also, they were as far advanced as the Egyptians. A work on astrology, prepared for Sargon, King of Babylon, almost 5000 years ago, includes a long list of the times of eclipses.

It is easy to understand why the sky watchers of those days were so interested in eclipses. Astrology was a strange mixture of science and magic. The priests claimed that, from their observation of the heavens, they could foretell all kinds of things — the outcome of battles, the fortunes of kings, the wrath of the gods. If the priest could forecast eclipses with accuracy, people were more ready to listen to other prophecies he might make and to give him more power.

We now know that an eclipse of the moon occurs when the earth is in a straight line between sun and moon. The earth then casts a shadow across the moon's face. The priests of Mesopotamia forecast eclipses of the moon with sufficient reliability to make us think that they probably knew this.

Looking up at the round edge of the shadow on the partly eclipsed moon, they would then realize that the earth itself must be round. Babylonian scribes did indeed draw fanciful maps of an earth whose shape was like a penny. On one such map, which archaeologists have found in recent times, Babylon occupies a large area in the middle.

Exercises:
1. Use the square-table method of multiplication to find 14×12.
2. If regular hexagons are drawn inside and outside a circle of a diameter 1, as in the drawing on page 112, each side of the inside hexagon is $\frac{1}{2}$, and each side of the outside hexagon is approximately .58. Use this information to find an approximate value of π. (See page 120 for the answers.)

Fermat's Method of Factoring

by Oystein Ore

*On page 110, square numbers were used to find the product of two
numbers. In the next selection they are used to do the opposite,
to find factors of a given number. The selection is from*
Number Theory and Its History *by Oystein Ore.*

This method is due to the French mathematician and lawyer Pierre
de Fermat (1601-1665), whose name we shall encounter repeatedly
in the following.

      Fermat must be awarded the honor of being the founding
father of number theory as a systematic science. His life was quiet
and uneventful and entirely centered around the town of Toulouse,
where he first studied jurisprudence, practiced law, and later became prominent as councilor of the local parliament. His leisure
time was devoted to scholarly pursuits and to a voluminous correspondence with contemporary mathematicians, many of whom,
like himself, were gentlemen-scholars, the ferment of intellectual
life in the seventeenth and eighteenth centuries. Fermat possessed
a broad knowledge of the classics, enjoyed literary studies, and
wrote verse, but mathematics was his real love. He published practically nothing personally, so that his works have been gleaned from
notes that were preserved after his death by his family, and from
letters and treatises that he sent to his correspondents. In spite of
his modesty, Fermat gained an outstanding reputation for his mathematical achievements. He made considerable contributions to the
foundation of the theory of probability in his correspondence with
Pascal and introduced coordinates independent of Descartes. The
French, when too exasperated over the eternal priority squabble
between the followers of Newton and Leibniz, often interject the
name of Fermat as a cofounder of the calculus. There is considerable

justification for this point of view. Fermat did not reduce his procedures to rule-of-thumb methods, but he did perform a great number of differentiations by tangent determinations and integrations by computations of numerous areas, and he actually gave methods for finding maxima and minima corresponding to those at present used in the differential calculus.

In spite of all these achievements, Fermat's real passion in mathematics was undoubtedly number theory. He returned to such problems in almost all his missives; he delighted to propose new and difficult problems, and to give solutions in large figures that require elaborate computations; and most important of all, he announced new principles and methods that have inspired all work in number theory after him.

Fermat's factorization method, which is the point interesting us particularly for the moment, is found in an undated letter of about 1643, probably addressed to Mersenne (1588–1648). Mersenne was a Franciscan friar and spent most of his lifetime in cloisters in Paris. He was an aggressive theologian and philosopher, a schoolmate and close friend of Descartes. He wrote some mathematical works, but a greater part of his importance in the history of mathematics rests on the fact that he was a favorite intermediary in the correspondence between the most prominent mathematicians of the times.

Fermat's method is based upon the following facts. If a number n can be written as the difference between two square numbers, one has the obvious factorization

$$n = x^2 - y^2 = (x - y)(x + y) \tag{4-2}$$

[The reason why equation (4–2) is correct is shown in the diagram on the next page. The number x^2 is represented by the area of a square whose side has length x. The number y^2 is represented by the area of a square whose side has length y. (Diagram (1).) Then $x^2 - y^2$ is the area that is left when a piece equal to the small square is removed from the large square. (Diagram (2).) In diagram (3) we see that this area is made up of two rectangles, I and II. In diagram (4) these rectangles have been rearranged to form a single rectangle whose length is $x + y$ and whose width is $x - y$, so that its area is $(x - y)(x + y)$. Consequently $x^2 - y^2 = (x - y)(x + y)$.]

On the other hand when

$$n = ab, b \geq a$$

is composite, one can obtain a representation (4–2) of n as the difference of two squares by putting

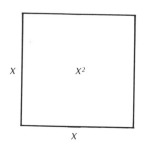

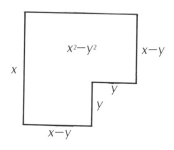

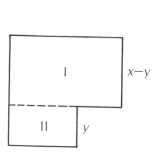

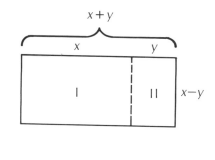

$$x - y = a, \qquad x + y = b$$

so that

$$x = \frac{b + a}{2}, \qquad y = \frac{b - a}{2} \qquad\qquad (4\text{-}3)$$

[For example, $35 = 5 \times 7$. If we write

$$x + y = 7, \text{ and}$$
$$x - y = 5,$$

adding the two equations gives us $2x = 7 + 5$, so

$$x = \frac{7 + 5}{2} = 6.$$

Subtracting the two equations gives us $2y = 7 - 5$, so

$$y = \frac{7 - 5}{2} = 1.$$

Then $35 = 5 \times 7 = (6 - 1)(6 + 1)$.]

Since we deal with the question of factoring n, we can assume that n ✱ is odd; hence a and b are odd and the values of x and y are integral.

✱ **If n is even, factoring it is no problem because we know that 2 is a factor.**

[If either a or b were even, their product would be even. So if the product, n, is odd, both a and b must be odd. When a and b are both odd, their sum and their difference are both even, so dividing the sum $a + b$ and the difference $a - b$ by 2 gives a whole number as a quotient.]

Corresponding to each factorization of n there exists, therefore, a representation (4–2). To determine the possible x and y in (4–2), we write

$$x^2 = n + y^2$$

Since $x^2 \geqq n$ one has $x \geqq \sqrt{n}$. The procedure consists in substituting successively for x the values above $\sqrt{n}$ and examining whether the corresponding

$$\Delta(x) = x^2 - n$$ ✱

is a square y^2. Let us illustrate by a simple example.

✱ **The symbol $\triangle$ (x) is read as delta of x, and it stands for the remainder that is left when n is subtracted from x^2.**

EXAMPLE

The number $n = 13{,}837$ is to be factored. One sees that $\sqrt{n}$ lies be- ✱ tween 117 and 118. In the first step we obtain

$$\Delta(118) = 118^2 - 13{,}837 = 87$$

✱ **That is, 13837 is between 13689 and 13924, and 13689 = 117², and 13924 = 118².**

which is not a square. In the next step one has

$$\Delta(119) = 119^2 - 13{,}837 = 324 = 18^2$$

so that we have found the factorization

$$13{,}837 = (119 - 18)(119 + 18) = 101 \cdot 137$$

This example is too simple to illustrate the short cuts that serve to facilitate the work with larger numbers. One important observation is that one need not calculate each $\Delta(x)$ separately. Since

$$(x + 1)^2 - n = x^2 - n + 2x + 1$$

one has

$$\Delta(x + 1) = \Delta(x) + 2x + 1$$

117

and by applying this rule repeatedly one finds
$$\Delta(x + 2) = \Delta(x + 1) + 2x + 3$$
$$\Delta(x + 3) = \Delta(x + 2) + 2x + 5$$
. . . .

This makes it possible to compute the successive Δ (x)'s by simple additions.

EXAMPLE

We shall take the formidable number $n = 2{,}027{,}651{,}281$, on which Fermat applied his method. The first integer above $\sqrt{n}$ is 45,030 and the calculations proceed as follows:

$x = 45{,}030$	$x^2 - n =$	49,619
	$2x + 1 =$	90,061
31		139,680
		90,063
32		229,743
		90,065
33		319,808
		90,067
34		409,875
		90,069
45,035		499,944
		90,071
36		590,015
		90,073
37		680,088
		90,075
38		770,163
		90,077
39		860,240
		90,079
45,040		950,319
		90,081
$x = 45{,}041$		$1{,}040{,}400 = 1{,}020^2 = y^2$

This shows that we have the factorization

$$n = (45{,}041 + 1{,}020)(45{,}041 - 1{,}020)$$
$$= 46{,}061 \cdot 44{,}021$$

where each factor can be shown to be a prime. In this chain of computations, each of the various numbers 49,619, 139,680, ... should be looked up in a table of squares* to determine whether it is actually a perfect square. However, in most cases this step may be eliminated since the last two digits will already show that the number is not a square. The table below giving the possible two last digits of a square number is most convenient for this purpose. Of all the numbers in the preceding chain it is only necessary to look up the numbers 499,944 and 1,040,400, since 44 and 00 may be the last two digits in a square.

Table of the last two digits in a square number				
00	21	41	64	89
01	24	44	69	96
04	25	49	76	
09	29	56	81	
16	36	61	84	

Fermat's method is particularly helpful when the number n has two factors whose difference

$$2y = b - a$$

is relatively small, because a suitable y will then quickly appear. In the choice of the example discussed above it is clear that Fermat had this in mind. By means of certain other improvements that can be introduced in the procedure, it becomes one of the most effective factorization methods available.

PROBLEM

Factor the following numbers by means of Fermat's method, using the table of squares on the next page. (a) 8927, (b) 8755, (c) 697, (d) 1679.
(Answers are on pages 120-121.)

*See page 120 for Table of Squares.

η	η^2	η	η^2	η	η^2	η	η^2
0	0	25	625	50	2 500	75	5 625
1	1	26	676	51	2 601	76	5 776
2	4	27	729	52	2 704	77	5 929
3	9	28	784	53	2 809	78	6 084
		29	841			79	6 241
4	16			54	2 916		
5	25	30	900	55	3 025	80	6 400
6	36	31	961	56	3 136	81	6 561
7	49	32	1 024	57	3 249	82	6 724
8	64	33	1 089	58	3 364	83	6 889
9	81	34	1 156	59	3 481	84	7 056
10	100	35	1 225	60	3 600	85	7 225
11	121	36	1 296	61	3 721	86	7 396
12	144	37	1 369	62	3 844	87	7 569
13	169	38	1 444	63	3 969	88	7 744
14	196	39	1 521	64	4 096	89	7 921
15	225	40	1 600	65	4 225	90	8 100
16	256	41	1 681	66	4 356	91	8 281
17	289	42	1 764	67	4 489	92	8 464
18	324	43	1 849	68	4 624	93	8 649
19	361	44	1 936	69	4 761	94	8 836
20	400	45	2 025	70	4 900	95	9 025
21	441	46	2 116	71	5 041	96	9 216
22	484	47	2 209	72	5 184	97	9 409
23	529	48	2 304	73	5 329	98	9 604
24	576	49	2 401	74	5 476	99	9 801
						100	10 000

Answers to exercises on page 113:

1. $\dfrac{14 + 12}{2} = 13.$ $\dfrac{14 - 12}{2} = 1.$

$13^2 - 1^2 = 169 - 1 = 168$

Therefore $14 \times 12 = 168.$

2. perimeter of inside hexagon $= 6 \times \frac{1}{2} = 3.$

perimeter of outside hexagon $= 6 \times .58 = 3.48.$

approximate value of $\pi = \dfrac{3 + 3.48}{2} = 3.24$

Answers to exercises on page 119:

(a) Let n = 8927. It is between 8836 and 9025, which occur in the table of squares, and $9025 = 95^2$.

If $x = 95,$ $x^2 - n = 98$

$2x + 1 = 191$

$\overline{}$

96          $\overline{289} = 17^2$

Use $x = 96,$ $y = 17.$

$8927 = (96 + 17)(96 - 17) = 113 \times 79.$

120

(b) Let $n = 8755$. It is between 8649 and 8836, which occur in the table of squares, and $8836 = 94^2$.

If $x = 94$, $\quad x^2 - n = 81 = 9^2$.
Use $x = 94$, $\quad y = 9$.
$8755 = (94 + 9)(94 - 9) = 103 \times 85$.

(c) Let $n = 697$. It is between 676 and 729, which occur in the table of squares, and $729 = 27^2$.

If $x = 27$, $\quad x^2 - n = 32$

$2x + 1 = \underline{55}$

28 $\qquad\qquad\qquad$ 87

$\qquad\qquad\qquad\quad \underline{57}$

29 $\qquad\qquad\qquad$ $144 = 12^2$.

Use $x = 29$, $\quad y = 12$.
$697 = (29 + 12)(29 - 12) = 41 \times 17$.

(d) Let $n = 1679$. It is between 1600 and 1681, which occur in the table of squares, and $1681 = 41^2$.

If $x = 41$, $\quad x^2 - n = \ ?$

$2x + 1 = \underline{83}$

42 $\qquad\qquad\qquad$ 85

$\qquad\qquad\qquad\quad \underline{85}$

43 $\qquad\qquad\qquad$ 170

$\qquad\qquad\qquad\quad \underline{87}$

44 $\qquad\qquad\qquad$ 257

$\qquad\qquad\qquad\quad \underline{89}$

45 $\qquad\qquad\qquad$ 346

$\qquad\qquad\qquad\quad \underline{91}$

46 $\qquad\qquad\qquad$ 437

$\qquad\qquad\qquad\quad \underline{93}$

47 $\qquad\qquad\qquad$ 530

$\qquad\qquad\qquad\quad \underline{95}$

48 $\qquad\qquad\qquad$ $625 = 25^2$.

Use $x = 48$, $\quad y = 25$.
$1679 = (48 + 25)(48 - 25) = 73 \times 23$.

PART 8

Earth, Air, and Water

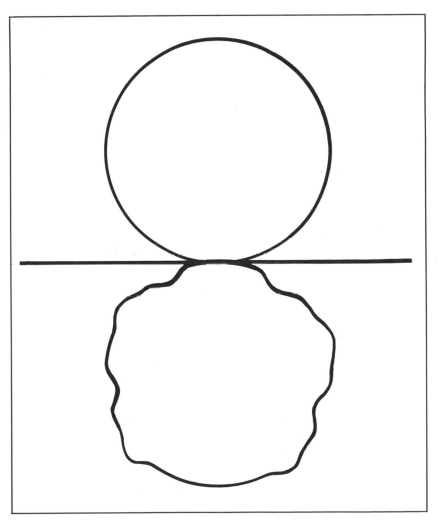

Isaac Newton "All Was Light"

by Isaac Asimov

*Isaac Newton was one of the great scientists of all time.
He discovered the law of gravitation, and, with his laws of
motion, put the science of mechanics on a firm foundation.
He invented a new branch of mathematics, now known as "the
calculus." He proved that sunlight is a mixture of colors, and
he invented the reflecting telescope. The story of Isaac Newton's
discoveries and inventions is told in the next selection, taken
from Isaac Asimov's* Breakthroughs in Science.

*Isaac Asimov is the author of more than one hundred
books of science information and science fiction.*

In 1666, so the story goes, when Isaac Newton was 23 years old,
he saw an apple fall from a tree. He had seen this happen before.
So had countless others. But this time Newton looked upward.
A pale half-moon was visible in the daytime sky over the English
countryside. Newton asked himself: Why does not the moon, too,
fall to the Earth, as the apple fell, drawn by the force of gravity?

Newton reasoned as follows: Perhaps the moon *is* pulled
to the Earth, but the speed of the moon's movement through
space cancels out the pull of the Earth's gravity. He reasoned
further: If the force that pulls the apple to the Earth also pulls
the moon to the Earth, that force would have to extend far out
into space. And as it extended into space, this force of gravity
would become weaker and weaker.

Newton calculated the distance of the moon from the
Earth's center. Then he calculated how fast the moon would have
to travel in its orbit to balance the pull of Earth's gravity at that
distance from the Earth. The answer he found checked pretty
nearly with the figures given by astronomers for the speed of the

moon. But it did not check *exactly*. The theory must be wrong, thought Newton. He put it aside.

Newton was already making his mark in mathematics, though as a youngster he had shown little promise. He was born on Christmas Day of 1642 (the year Galileo died) in Woolsthorpe, England. His father, a farmer, had died the day before young Isaac was born. As a boy, Newton was a dull student until (the story goes) he grew tired of being beaten up by the bright boy of the class. Newton applied himself until *he* was first in the class.

By the time he was 18, Newton's interest in mathematics was arousing attention. He would make a poor farmer, said his uncle, and he talked Newton's mother into sending the young man to Cambridge University. Nine years later, Newton was professor of mathematics at Cambridge.

But what years these were for Newton! He studied beams of light, for instance. He allowed sunlight to enter a darkened room through a hole in a curtain. The tiny beam of light then was passed through a triangular glass prism. The light fell on a screen as a rainbow band — not as a spot of white light. Newton was the first to discover that white light was actually made up of various colors which could be separated and recombined.

At the same time, he blazed new frontiers in mathematics. He worked out the binomial theorem for expressing cer- ✳ tain algebraic quantities. Much more important, he discovered a new way for calculating areas bounded by curves. (At almost the same time, the German mathematician Wilhelm Leibniz discovered this independently.) Newton called his new technique "fluxions." We call it "calculus."

✳ **The binomial theorem is a general rule sttudied in algebra for calculating the value of a power of a sum, $(a + b)^n$.**

Even Newton's mistakes were fruitful in their results. Newton had developed a theory to explain his discovery that white light could be bent by glass into a rainbow. The theory was wrong, as scientists found later. But it seemed to explain why the early telescopes, which were constructed of lenses that bent or "refracted" light, formed images surrounded by small colored blurs. This was called "chromatic aberration." Newton's wrong theory led him to believe that this chromatic aberration could never bc corrected.

For this reason, he decided to make telescopes without

lenses. He designed telescopes that used parabolic mirrors to gather and concentrate light by reflection. He built his first in 1668. These "reflecting telescopes" had no chromatic aberration.

Shortly after Newton died, telescopes were built using special lenses that did *not* show chromatic aberration. However, the best and largest telescopes still use the reflecting principle. The 200-inch telescope atop Mt. Palomar in California is a reflecting telescope.

But Newton's attempt to apply Earth's gravity to the moon remained a failure. The years passed, and it seemed dead for good.

As it happened, one of Newton's faults was that he couldn't take criticism, and he kept up feuds. Newton and his followers, for instance, fought a battle with Leibniz and his followers over who first invented calculus, when both deserved credit.

In the Royal Society of London (of which Newton was a member), Newton's great enemy was Robert Hooke. Hooke was a capable scientist, but he had a grasshopper mind. He started things and dropped them. He started so many things that no matter what anyone else did, Hooke could always claim he had thought of it first.

In 1684, Hooke — in the company of Edmund Halley, a very good friend of Newton's — boasted that he had worked out the laws explaining the force that controlled the motions of the heavenly bodies. His theory didn't seem satisfactory, and arguments began.

Halley went to Newton and asked him how the planets would move if there were a force of attraction between bodies that weakened as the square of the distance.

Newton said at once, "In ellipses."

"But how do you know?"

"Why, I have calculated it." And Newton told his friend the story of his attempt 18 years earlier, and how it had failed. Halley, in a frenzy of excitement, urged Newton to try again.

Now things were different. In 1666, Newton had *supposed* that the force of attraction acted from Earth's center, but he hadn't been able to prove it. Now he had calculus as a tool. With his new mathematical techniques, he could *prove* that the force acted from the center. Also, during the last 18 years, new and better measurements had been made of the radius of the Earth, and of the moon's size and its distance from the Earth.

This time, Newton's theory checked the facts — exactly. The moon was pulled to the Earth, held in Earth's grip by gravity, just as the apple.

In 1687, Newton expounded his theory in a book, *Philosophiae Naturalis Principia Mathematica*. In it, he also announced the "Three Laws of Motion." The third of these states that for every action there is an equal and opposite reaction. It is this principle which explains how rocket engines work.

The Royal Society intended to publish the book, but there was not enough money in the treasury. Also, Hooke was raising all the trouble he could, insisting he had the idea first. Halley, who was well-to-do, therefore published the book at his own expense.

But the great days of Newton were over. In 1692, that all-embracing mind tottered. Newton had a nervous breakdown and had to spend nearly two years in retirement. To burn up his boundless mental energies, he turned toward theology and alchemy, as though science were not enough. He wasted his powers on a search for ways to manufacture gold.

He was never the same after his nervous breakdown, though at times he showed flashes of his old genius. In 1696, for example, a Swiss mathematician challenged Europe's scholars to solve two problems. The day after Newton saw the problems he forwarded the solutions anonymously. The Swiss mathematician penetrated the disguise at once. "I recognize the claw of the lion," he said.

Newton was appointed Warden of the Mint in 1696, and placed in charge of coinage. He resigned his professorship to attend to his new duties. These he did so well that he was a virtual terror to counterfeiters.

He even served in Parliament for two terms, elected as a representative of Cambridge University. He never made a speech. On one occasion he rose, and the House fell silent to hear the great man. All Newton did was ask that the window be closed because there was a draft.

In 1705, Newton was knighted by Queen Anne. On March 20, 1727, forty years after his great discoveries, Newton died.

Newton is important for more than his great discoveries. To be sure, his laws of motion completed the work begun by Galileo. And his laws of universal gravity explained the work of Copernicus and Kepler, as well as the movement of the tides.

These great concepts live today in every branch of mechanics. He founded the science of optics, which enabled us to learn as much as we have about the composition of the stars and almost all we have learned about the composition of matter. The value of calculus in every branch of science is beyond estimate.

Yet Newton's greatest importance to the advance of science may be psychological. The reputation of the ancient Greek philosophers and scholars had been badly shaken by the discoveries of such moderns as Galileo and Harvey. But Europe's scientists still suffered a sense of inferiority.

Then came Newton. His gravitational theories opened a vision of the universe greater and grander than anything Aristotle had dreamed of. His elegant system of celestial mechanics brought the heavens within the reach of man's intelligence, showing that the most remote heavenly bodies were subject to precisely the same laws as the smallest mundane object.

His theories became models of what a scientific theory should be. In all other sciences, and in political and moral philosophy also, writers and thinkers since Newton have attempted to emulate his elegant simplicity. They used rigorous formulae and a few basic principles.

Here was a mind as great as any of the ancients. His contemporaries knew it. Newton was almost idolized in his own lifetime. When he died, he was buried in Westminster Abbey with England's heroes. Voltaire of France, who was visiting England at the time, commented with admiration that England honored a mathematician as other nations honor a king.

From Newton's day, science has been filled with a self-confidence that never again faltered.

Newton's glory is perhaps best expressed in a couplet by Alexander Pope.

Nature and Nature's laws lay hid in night.
God said, Let Newton be! and all was light.

The Law of Gravitation

Newton's Law of Gravitation says that every particle of mass exerts a pull on every other particle of mass. The strength of the pull varies as the amount of the mass. That is, if the mass is multiplied by 2, the pull is multiplied by 2; if the mass is multiplied by 3, the pull is multiplied by 3; etc. The strength of the pull of one mass on another also varies inversely as the distance between them. That means that if you multiply the distance by 2, the pull is divided by 2^2 or 4; if you multiply the distance by 3, the pull is divided by 3^2 or 9; if you multiply the distance by m, the pull is divided by m^2.

If a mass occupies a very small space, Newton assumed that it is concentrated at a point. Then the distance between two such masses is the distance between the points at which the masses are concentrated, and the pull between the masses is easily calculated. However, a mass like that of the earth is not concentrated at a point. It is spread out throughout a sphere whose radius is about 8000 miles. How do you calculate the pull exerted by such a spherical mass? To answer this question, Newton imagined a spherical mass to be divided into thin shells, like the layers of an onion, and imagined each shell divided into many small particles. Then each particle exerts its own pull, and the total pull of the spherical mass is the sum of all the pulls of all the particles in all the shells.

Newton found that he could easily add the pulls of the particles if he considered first only one thin shell at a time. But then he had to consider two problems separately: one problem was to calculate the pull of a shell on a particle of mass that is inside the hollow space in the shell. The other problem was to calculate the pull of a shell on a particle of mass that is outside the shell. The next selection is Newton's solution to the first problem. It is taken from his book, *Mathematical Principles of Natural Philosophy*.

In Newton's discussion, he thinks of the thin spherical shell as being so thin that it is like a *spherical surface*. He refers to the particle of mass inside the hollow of the shell as a *corpuscle*, which means *small body*. He calls the force exerted by a particle as a *centripetal force*, which means that it is a pull toward the particle.

The Pull of a Shell on a Particle Inside

by Isaac Newton

If to every point of a spherical surface there tend equal centripetal forces decreasing as the square of the distances from those points, I say, that a corpuscle placed within that surface will not be attracted by those forces any way.

Let HIKL be that spherical surface, and P a corpuscle placed within. Through P let there be drawn to this surface two lines HK, IL, intercepting very small arcs HI, KL; and because (by Cor. III, Lem. VII) the triangles HPI, LPK are alike, those arcs will be proportional to the distances HP, LP; and any particles at HI and KL of the spherical surface, terminated by right lines passing through P, will be as the square of those distances.

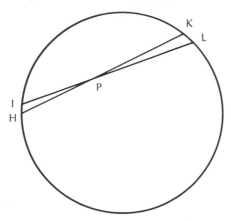

[Newton's argument so far is based on the following ideas:

1. The arcs HI and KL are so small that we may think of them as being practically like small straight line segments.
2. In triangle PIH and triangle PKL, the angles at the point P are equal to each other. The angle at I is equal to the angle at K, and the angle at H is equal to the angle at L. (These facts are proved in geometry.) So the triangles have the same shape, and consequently their corresponding sides are proportional. That is, if LP is m times as long as HP, then KL is also m times as long as HI.
3. Think of the particle at HI as covering a small area; then the particle at KL covers an area with the same shape, as shown in the diagram below. Since KL is m times as long as HI, the area

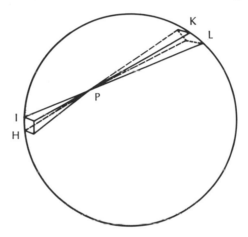

at KL is m^2 times as great as the area at HI. (See page 129.) Since we are assuming that the mass is spread out uniformly over the surface of the sphere, the mass in the area at KL is m^2 times as great as the mass in the area at HI.]

Therefore the forces of these particles exerted upon the body P are equal between themselves. For the forces are directly as the particles, and inversely as the square of the distances. And these two ratios compose the ratio of equality, 1 : 1.

[Newton arrives at his conclusion in this way: Let s be the strength of the pull exerted by the mass at HI. To calculate in terms of s the strength of the pull exerted by the mass at KL, we must take into account two things: because the mass at KL is m^2 times as great as the mass at HI we must *multiply* s by m^2; because the distance LP is *m* times as great as the distance HP, we must *divide* s by m^2. But *multiplying* by m^2 and *dividing* by m^2 cancel each other out. So the pull exerted by the mass at KL is exactly s. That is, the *mass at KL and the mass at HI exert equal pulls in opposite directions.]*

The attractions therefore, being equal, but exerted in opposite directions, destroy each other. And by a like reasoning all the attractions through the whole spherical surface are destroyed by contrary attractions. Therefore the body P will not be any way impelled by those attractions. Q.E.D.

[Thus Newton was able to show that a shell exerts no force on a particle inside the hollow of the shell. In a separate proof, he showed that a shell does pull on a particle that is outside the shell, and it does so as if all the mass of the shell were concentrated at its center.]

Kepler's Second Law

In the year 1609, Johannes Kepler published two laws of planetary motion that he had discovered by analysing the measurements of planetary motion made by the astronomer Tycho Brahe. The first law says that the orbit of a planet is an ellipse. The second law says that the line joining a planet to the sun sweeps over equal areas in equal times. Isaac Newton later showed that both laws are consequences of the fact that a planet moves under the influence of the force of gravitation tending to pull it toward the sun. The next selection is Newton's proof that if a body moves under the influence of a force that is directed toward a fixed point, the line that joins the body to that fixed point sweeps over equal areas in equal times.

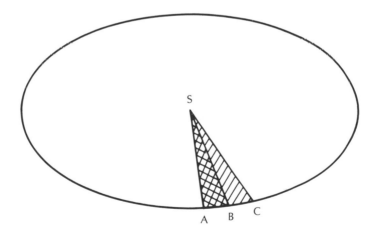

Kepler's Second Law: If a planet moves from A to B and then from B to C in equal intervals of time, then the area between SA and SB equals the area between SB and SC.

The Influence of a Central Force

by Isaac Newton

[Newton's text begins on page 136. To follow Newton's argument more easily, let us first examine it one step at a time, and build up his diagram step by step.

Newton begins by examining the motion of the body during a very short interval of time. During that time it travels over only a small part of its path. Even though the path may be curved, each small part of it is pretty close to being a straight line. So Newton uses the straight line (right line) AB as an approximate picture of the path of the body during a short interval of time.

He considers next another short interval of time equal to the first one, and examines how the body will move during this second interval. He does this in two steps. First he figures out how the body would move if there were no force pulling it toward S. Then he figures out what correction must be made to take into account the fact that the body is being pulled by a force directed toward S.

If the body were not being pulled by a force directed toward S, its motion would be based on Newton's First Law, which says that a body that is not acted on by a force continues to move in a straight line at the same speed that it had to begin with. But motion at an unchanging speed will carry it across *equal distances in equal times*. So Bc, the distance the body moves during the second interval of time, is equal to AB, the distance it moved during the first interval of time. This part of the diagram is reproduced below:

134

We have also put into the diagram the line ST drawn perpendicular to AB. In triangles SAB and SBc, let us use AB and Bc as the bases of the triangles respectively. The line segment ST is the altitude drawn to the side AB in triangle SAB. The same line ST is also the altitude drawn to the side Bc in triangle SBc. The triangles SAB and SBc thus have equal bases and also equal altitudes. Consequently, they have equal areas.

Now, to make the correction based on the fact that the body is being pulled by a force directed toward S, Newton uses this rule based on his laws of motion: When two influences act on the motion of a body at the same time the change in position that results is the same as if they had acted separately one after the other. In this case, to find where the body will be after the second interval of time if, while it is at B, a force pulls it in the direction BS, we consider separately the two influences on the body, namely, its original speed, and the pull toward S in the direction BS. We know already that its original speed alone would carry it from B to c. During the same interval of time, if only the pull toward S were acting, the body would have moved along the line BS to some point V. If we think of these two influences as separated and acting one after the other, then we picture the motion in the direction BS as taking place after the body has reached the point c. That is, the influence of the pull toward S would be to carry the body a distance cC equal in length to BV and parallel to the direction BS. Thus the combined effect of the two influences is to move the body from B to C along the line BC, where C and c are on a line parallel to BS. (See the diagram below.)

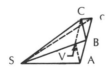

Then the area swept over during the second interval by the line joining the body to S is the area of the triangle SBC. Let us compare this area with the area of the triangle SBc. To do so, draw CQ and cR perpendicular to the line SB.

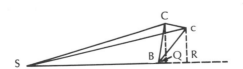

Since Cc is parallel to SB, and parallel lines are the same distance apart at all points, CQ is equal in length to cR. In triangle SBc, if we use SB as base, then cR is the altitude to that base. In triangle SBC, if we use SB as base, then CQ is the altitude to that base. Consequently triangles SBC and SBc have equal bases (SB) and equal altitude (CQ and cR), so they have equal areas.

Since we have shown that triangles SBC and SAB are both equal in area to triangle SBc, they are equal in area to each other.

The rest of Newton's proof repeats the argument to extend it to a third interval of time, a fourth interval of time, and so on. Then he makes the intervals smaller and smaller to eliminate the small error that may have crept in when he assumed that the body moves in a straight line during a small interval of time.

Now let us see how Newton developed this argument in his own words.]

The areas which revolving bodies describe by radii drawn to an immovable centre of force do lie in the same immovable planes, and are proportional to the times in which they are described.

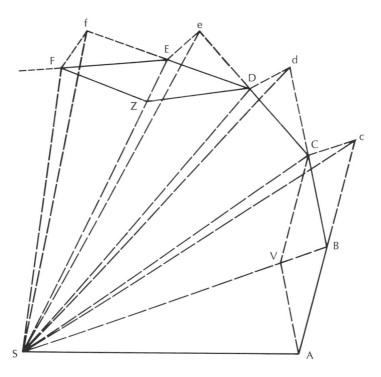

For suppose the time to be divided into equal parts, and in the first part of that time let the body by its innate force describe the right line AB. In the second part of that time, the same would (by Law 1), if not hindered, proceed directly to c, along the line Bc equal to AB; so that by the radii AS, BS, cS, drawn to the centre, the equal areas ASB, BSc, would be described. But when the body is arrived at B, suppose that a centripetal force acts at once with a great impulse, and, turning aside the body from the right line Bc compels it afterwards to continue its motion along the right line BC. Draw cC parallel to BS, meeting BC in C; and at the end of the second part of the time, the body (by Cor. 1 of the Laws) will be found in C, in the same plane with the triangle ASB. Join SC, and, because SB and Cc are parallel, the triangle SBC will be equal to the triangle SBc, and therefore also to the triangle SAB. By the like argument, if the centripetal force acts successively in C, D, E, &c., and makes the body, in each single particle of time, to describe the right lines CD, DE, EF, &c., they will all lie in the same plane; and the triangle SCD will be equal to the triangle SBC, and SDE to SCD, and SEF to SDE. And therefore, in equal times, equal areas are described in one immovable plane: and, by composition, any sums SADS, SAFS, of those areas, are to each other as the times in which they are described. Now let the number of those triangles be augmented, and their breadth diminished *in infinitum*; and (by Cor. IV, Lem. III) their ultimate perimeter ADF will be a curved line: and therefore the centripetal force, by which the body is continually drawn back from the tangent of this curve, will act continually; and any described areas SADS, SAFS, which are always proportional to the times of description, will, in this case also, be proportional to those times.

Q.E.D.

Using Arithmetic, Algebra, and Geometry to Weigh the Air Around the Earth

by Hy Ruchlis and Jack Engelhardt

*The air that surrounds the earth reaches up several hundred miles
above the ground. Yet you can weigh the air around the earth
without leaving your desk. Hy Ruchlis and Jack Englehardt show
you how in the next selection, taken from their book*
The Story of Mathematics.

Anybody who works with his mind can use mathematics as a basic
tool. This is obviously true of engineers, scientists, and accountants.
But it is also true of lawyers, doctors, salesmen, teachers, detectives,
and people in general. To show the great power of mathematics,
let's use it to solve a problem that seems impossibly difficult.

What is the approximate weight of the air on earth?

This may look like a puzzle that has no answer. But with
some mathematics it is not difficult to solve.

You are probably familiar with a mercury barometer. This
device (Fig. 153) is used to measure air pressure. Mercury is a very
heavy liquid. It is placed in a curved tube that has no air in it (a
vacuum) at the closed top (A in Fig. 153). The shorter end (B in
Fig. 153) is open to the air.

It would seem that the mercury would spill out of the
open end. But it doesn't. Air presses in at the open end of the tube
and pushes the mercury up into the other side of the tube. If the
pressure of the air is less, it doesn't push so hard and the level of
the mercury drops. That's how the barometer is used to measure
the air pressure.

With this simple device, scientists have found that the air pressure is about 15 pounds per square inch on the surface of the earth.

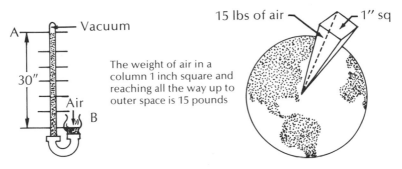

15 lbs of air — 1" sq

A — Vacuum

30"

Air B

The weight of air in a column 1 inch square and reaching all the way up to outer space is 15 pounds

A barometer measures air pressure.

What causes air pressure? The air above us has weight. When the barometer shows a pressure of 15 pounds per square inch it means that all of the air in a column 1 inch thick (1 square inch) all the way up to outer space, weighs 15 pounds (Fig. 154).

If we know how many square inches there are on the earth's surface, we can easily figure out the total weight of all the air on earth by simple multiplication. *But we need not measure the number of square inches on the earth. We can calculate it from a formula.* The only fact we must know is the diameter (D) of the earth.

Mathematicians give us this formula for the area of a sphere:

$$A = \pi \times D^2$$

If you are not familiar with this formula, it may look like nonsense. Yet, if you know something about the language of algebra, it is easy to use.

In the formula above, A stands for "area." The symbol π is the Greek letter "pi" (pronounced like pie). This letter stands for how many times longer the outside (circumference) of a circle is, than its diameter.

The distance around the outside (circumference) of any circle is about $3\frac{1}{7}$ times longer than the diameter. If we need greater accuracy, we can use the number 3.14159. If this is not accurate enough, we can carry it out to additional decimal places as follows:

3.141 592 653 589 793 238 462 643 383 279 502 88 . . .

139

Mathematicians have figured out this number to more than 700 decimal places. These decimal places can go on forever, because π is the kind of number that is impossible to express with a definite number of digits. For most purposes $3\frac{1}{7}$ (or $\frac{22}{7}$) is close enough. Sometimes, for a rough approximation, the number 3 is sufficient.

"D" represents the diameter of the sphere. D^2 means that D is to be multiplied by itself ($D \times D$). Read D^2 as "D squared."

We use the name "squared" because to find the area (surface) of a square we must multiply one side of the square by itself. For example, the area of a square 3 inches long on each side is 3×3 or 9 square inches.

What would D^3 mean? It means D multiplied by itself 3 times ($D \times D \times D$). We speak of it as "D cubed" because we get the volume (space occupied by) of a cube by multiplying its side by itself 3 times.

Now let's go to work with our formula.

$$A = \pi \times D^2$$

Because D^2 means $D \times D$, we can also write our formula as:

$$A = \pi \times D \times D$$

You know that the diameter of the earth is 8,000 miles in length. Shall we put the number 8,000 in place of D and multiply to find the area of the earth? Yes, but we must be careful about one thing. If we use 8,000 miles for D, the answer will come out in square miles. We want the area in square inches. We must therefore change the 8,000 miles into feet and then into inches. We know that there are 5,280 feet to a mile and 12 inches to a foot. So we multiply 8,000 by 5,280 and then by 12 to find the number of inches in 8,000 miles.

Let's put all our information into the formula.

$$A = \pi \times D \times D$$
$$= \tfrac{22}{7} \times (8{,}000 \times 5{,}280 \times 12) \times (8{,}000 \times 5{,}280 \times 12)$$
square inches.

When this multiplication is carried out we will have the number of square inches on the surface of the earth! This number must then be multiplied by 15 pounds per square inch to get the weight of all the air on earth. It is a good idea to put all numbers to place before any multiplying is done. So, we get:

140

Weight of air on earth (in pounds) =

$$15 \times \frac{22}{7} \times (8{,}000 \times 5{,}280 \times 12) \times (8{,}000 \times 5{,}280 \times 12).$$

Here is an important mathematical rule to remember at this point: when you have several numbers to multiply, it makes no difference in the answer which numbers you multiply first. Any order will do. For example, $2 \times 3 \times 4$ is 24, just the same as $4 \times 2 \times 3$ equals 24.

We may, therefore, arrange the numbers in our big multiplication in any order we like. Let's save some work by choosing a better order for the numbers.

We may multiply the numbers exactly, if we wish. But it will be quite a job. So, we'll approximate and get an answer that is close to the actual multiplication. We do this by rounding off the numbers. Thus we can approximate 22/7 by using 3. The number 5,280 may be approximated by using 5,000.

We get:

Weight of air on earth (in pounds) =

$$15 \times 3 \times (8{,}000 \times 5{,}000 \times 12) \times (8{,}000 \times 5{,}000 \times 12) \text{ lbs.}$$

Now, we pick easy combinations of numbers for multiplying. To avoid carrying along a lot of zeros when we multiply, we shift them all into one number at the right. Remember that when we cross off a zero at the end of one whole number, we must make up for it by attaching it to the end of another whole number. When we cross off a zero, we divide by 10. When we attach a zero to the end of another number, we multiply by 10. The net result of doing both in a multiplication (crossing off a zero in one place, and attaching one elsewhere) gives us the same answer. If there is no number to which the zero is to be attached, we start with the number 1 and attach zeros to that.

Weight of air (in pounds) =

$$15 \times 3 \times (8{,}000 \times 5{,}000 \times 12) \times (8{,}000 \times 5{,}000 \times 12)$$

Crossing off the 12 zeros and tacking them on at the end gives us

$$= (15 \times 3) \times (8 \times 5 \times 12) \times (8 \times 5 \times 12) \times 1{,}000{,}000{,}000{,}000$$

Multiplying:

$$= (45) \times (480) \times (480) \times 1{,}000{,}000{,}000{,}000 \text{ lbs.}$$

Now we can get an approximate answer by lowering 45 to 40 and raising 480 to 500.

$$\text{Weight of air} = 40 \times 500 \times 500 \times 1{,}000{,}000{,}000{,}000$$
$$= (4 \times 5 \times 5) \times 100{,}000{,}000{,}000{,}000{,}000{,}000$$
$$= 100 \times 100{,}000{,}000{,}000{,}000{,}000{,}000$$
$$= 10{,}000{,}000{,}000{,}000{,}000{,}000{,}000 \text{ lbs.}$$

This is a very clumsy number. A mathematician would simplify it by writing 10^{19}. The complete reason for the choice of 10 in this number takes us into mathematics beyond the level of this book. You can think of 10^{19} as meaning "tack on 19 zeros to the number 1."

Large numbers are read as follows:

1,000 . . . . . one thousand

1,000,000 . . . . . one million

1,000,000,000 . . . . . one billion (but not in England, where a billion is equal to our trillion).

1,000,000,000,000 . . . . . one trillion

1,000,000,000,000,000 . . . . . one quadrillion

1,000,000,000,000,000,000 . . . . . one quintillion

So, the weight of the air on earth is about 10 quintillion pounds. Or, if you prefer, the weight of the air is 10^{19} pounds. This expression is read as "ten to the nineteenth" pounds.

You may argue that this number is not exact because we raised some numbers and lowered others. You are right. But it makes little difference in the final result because the numbers we used were approximate, too.

The earth's diameter is not exactly 8,000 miles. The earth is not a perfect sphere. It is 27 miles less in diameter through the poles than through the equator.

The number 15 is not exact for air pressure. It is closer to 14.7 pounds per square inch at sea level. The pressure changes from day to day and is different all over the earth. On mountain tops the pressure drops greatly.

So, if our original figures are approximate our final answer must also be approximate, no matter how accurately we may multiply the numbers. However, scientists do this type of approximating quite often when they have to make complicated calculations with numbers that are not exact to start with.

At any rate, we have a pretty good idea about the weight of the air on earth, which we didn't have before.

How did we get this answer? We used certain kinds of knowledge.

142

We had to know:

Geometry — the mathematics of shapes. Knowledge in this field gave us the formula for the area of the outside surface of a sphere.

Algebra — this kind of mathematics tells us how to handle a formula properly to get the right answers.

Arithmetic — knowledge of numbers and how to work with them. This kind of mathematics enables us to calculate the answer when numbers are put into the formula.

Measurement — we needed this information to change miles into feet and then into inches.

Physics — the science of forces, motion and energy. We needed this knowledge to use the barometer and measure the air pressure.

Common sense and **Logic** — we needed these abilities to know when to combine numbers, how to make approximations, and how to juggle the numbers for ease in multiplication.

Do you see why we call mathematics a basic **tool**? Could you ever dream of weighing the air on the earth without it? Yet with this wonderful kind of knowledge you can sit down at a desk, and solve such a problem in a few minutes!

The Sea Monster

by Jules Verne

Just as the air around the earth presses down on everything that is on the ground, the water in the sea presses down on everything that is at the bottom of the sea. In the next selection, the French science-fiction writer Jules Verne uses this fact to try to prove that animals that live at the bottom of the sea must be very strong.

The selection is taken from Twenty Thousand Leagues Under the Sea, *first published in 1870. In this book Verne described an undersea voyage by a submarine long before any real submarine was built. In this respect his fiction foretold the truth that was yet to come. However, Verne was an imaginative writer and not a scientist, so he did make some mistakes in his forecasts. In fact, the conclusion that he leads to in this selection is wrong, and he seems to prove it only because he makes a false assumption. See if you can identify the false assumption as you read his argument. If you don't spot it, Rachel Carson, who was a scientist, will point it out to you in the very next selection taken from her book* The Sea Around Us.

Now, what was Ned Land's opinion upon the question of the marine monster? I must admit that he did not believe in the unicorn, and was the only one on board who did not share that universal conviction. He even avoided the subject, which I one day thought it my duty to press upon him. One magnificent evening, July the thirtieth — that is to say, three weeks after our departure — the frigate was abreast of Cape Blanc, thirty miles to leeward of the coast of Patagonia. We had crossed the tropic of Capricorn, and the Strait of Magellan opened less than seven hundred miles to the south. Before eight days were over, the *Abraham Lincoln* would be plowing the waters of the Pacific.

144

Seated on the poop, Ned Land and I were chatting of one thing and another as we looked at this mysterious sea, whose great depths had up to this time been inaccessible to the eye of man. I naturally led up the conversation to the giant unicorn, and examined the various chances of success or failure of the expedition. But seeing that Ned Land let me speak without saying too much himself, I pressed him more closely.

"Well, Ned," said I, "is it possible that you are not convinced of the existence of this cetacean that we are following? Have you any particular reason for being so incredulous?"

The bathyscaphe, a submarine laboratory used by the United States Navy for deep sea exploration.

The harpooner looked at me fixedly for some moments before answering, struck his broad forehead with his hand (a habit of his), as if to collect himself, and said at last, "Perhaps I have, Mr. Aronnax."

"But, Ned, you, a whaler by profession, familiarized with all the great marine mammalia; you, whose imagination might easily accept the hypothesis of enormous cetaceans, *you* ought to be the last to doubt under such circumstances!"

"That is just what deceives you, Professor," replied Ned. "That the vulgar should believe in extraordinary comets traversing space, and in the existence of antediluvian monsters in the heart of the globe, may well be; but neither astronomer nor geologist believes in such chimeras. As a whaler I have followed many a cetacean, harpooned a great number, and killed several; but, however strong or well-armed they may have been, neither their tails nor their weapons would have been able even to scratch the iron plates of a steamer."

"But, Ned, they tell of ships which the tusk of the narwhal has pierced through and through."

"Wooden ships — that is possible," replied the Canadian; "but I have never seen it done; and, until further proof, I deny that whales, cetaceans, or sea unicorns could ever produce the effect you describe."

"Well, Ned, I repeat it with a conviction resting on the logic of facts. I believe in the existence of a mammal powerfully organized, belonging to the branch of vertebrata, like the whales, the cachalots, or the dolphins, and furnished with a horn of defense of great penetrating power."

"Hum!" said the harpooner, shaking his head with the air of a man who would not be convinced.

"Notice one thing, my worthy Canadian," I resumed. "If such an animal is in existence, if it inhabits the depths of the ocean, if it frequents the strata lying miles below the surface of the water, it must necessarily possess an organization the strength of which would defy all comparison."

"And why this powerful organization?" demanded Ned.

"Because it requires incalculable strength to keep one's self in these strata and resist their pressure. Listen to me. Let us admit that the pressure of the atmosphere is represented by the weight of a column of water 32 feet high. In reality the column of water would be shorter, as we are speaking of sea water, the density of which is greater than that of fresh water. Very well, when you dive, Ned, as many times 32 feet of water as there are above you, so many times does your body bear a pressure equal to that of the atmosphere, that is to say, 15 pounds for each square inch of its surface. It follows then, that at 320 feet this pressure equals that of 10 atmospheres, of 100 atmospheres at 3,200 feet, and of 1,000 atmospheres at 32,000 feet; that is, about 6 miles; which is equivalent to saying that, if you could attain this depth in the ocean, each square three-eighths of an inch of the surface of your body would bear a pressure of 5,600 pounds. Ah! my brave Ned, do you know how many square inches you carry on the surface of your body?"

"I have no idea, Mr. Arronax."

"About 6,500; and, as in reality the atmospheric pressure is about 15 pounds to the square inch, your 6,500 square inches bear at this moment a pressure of 97,500 pounds."

"Without my perceiving it?"

"Without your perceiving it. And if you are not crushed

146

by such a pressure, it is because the air penetrates the interior of your body with equal pressure. Hence, perfect equilibrium between the interior and exterior pressure, which thus neutralize each other, and which allows you to bear it without inconvenience. But in the water it is another thing."

> ✱ Verne's or his translator's second use of the word <u>pressure</u> here is wrong. A pressure of 15 lbs. per square inch applied to an area of 6500 square inches exerts a <u>force</u> of 97,500 pounds.

"Yes, I understand," replied Ned, becoming more attentive; "because the water surrounds me, but does not penetrate."

"Precisely, Ned: so that at 32 feet beneath the surface of the sea you would undergo a pressure of 97,500 pounds; at 320 feet, ten times that pressure; at 3,200 feet, a hundred times that pressure; lastly, at 32,000 feet, a thousand times that pressure would be 97,500,000 pounds; that is to say, that you would be flattened as if you had been drawn from the plates of a hydraulic machine!"

"The devil!" exclaimed Ned.

"Very well, my worthy harpooner, if some vertebrate, several hundred yards long, and large in proportion, can maintain itself in such depths, of those whose surface is represented by millions of square inches, that is by tens of millions of pounds, we must estimate the pressure they undergo. Consider, then, what must be the resistance of their bony structure, and the strength of their organization to withstand such pressure!"

"Why!" exclaimed Ned land, "they must be made of iron plates eight inches thick, like the armored frigates."

"As you say, Ned. And think what destruction such a mass would cause, if hurled with the speed of an express train against the hull of a vessel."

"Yes — certainly — perhaps," replied the Canadian, shaken by these figures, but not yet willing to give in.

"Well, have I convinced you?"

"You have convinced me of one thing, sir, which is that, if such animals do exist at the bottom of the seas, they must necessarily be as strong as you say."

"But if they do not exist, mine obstinate harpooner, how explain the accident to the *Scotia*?"

Fragile Animals Under High Pressure

by Rachel Carson

At first thought it seems a paradox that creatures of such great fragility as the glass sponge and the jellyfish can live under the conditions of immense pressure that prevail in deep water. And the miracle is heightened when we recall a few simple facts about pressure in the sea. At sea level, the pressure of air on our bodies is about 1 atmosphere, or 15 pounds to the square inch of surface. Descending into the water, pressure increases by 1 atmosphere for every 33 feet. At the limit of diving-helmet range, there is a pressure of about 45 pounds on every square inch of a man's body. This is about as much as the unprotected human body can endure — a variation of 3 atmospheres. For creatures at home in the deep sea, however, the saving fact is that the pressure inside their tissues is the same as that without, and, as long as this balance is preserved, they are no more inconvenienced by a pressure of a ton or so than we are by ordinary atmospheric pressure. And most abyssal creatures, it must be remembered, live out their whole lives in a comparatively restricted zone, and are never required to adjust themselves to extreme changes of pressure.

PART 9

Rhythms

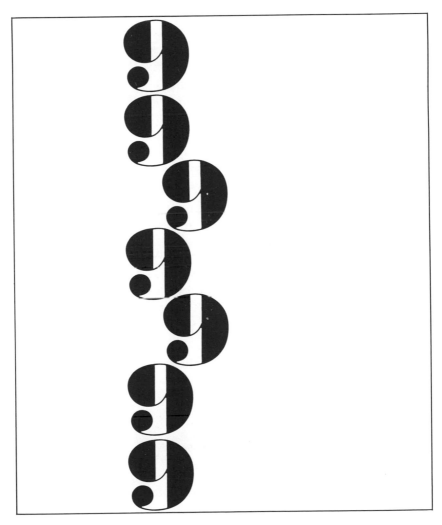

The Pendulum and Galileo

by Lancelot Hogben

A pendulum swings back and forth with a regular rhythm that depends only on the length of the pendulum. This fact was first discovered in 1583 by Galileo. We learn about this discovery and some of its consequences in the next selection, taken from The Wonderful World of Mathematics.

The first clue to accurate measurement of small intervals of time was discovered in 1583, when Galileo, a young Italian medical student, watched a lamp swinging to and fro in Pisa Cathedral. Timing its motion by the beat of his pulse, Galileo found that all swings, whether wide or narrow, took the same time.

Later on, when Galileo gave up the study of medicine to take up mathematics and physics, he used a home-made water clock to check the accuracy of this observation. While a pendulum

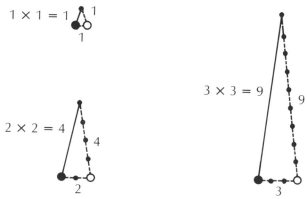

Each pendulum's length determines its time of swing

150

Galileo's design for a pendulum clock.

was swinging, he allowed water to flow from a hole at the bottom of a large vessel and fall into a small one below it. If the weight of water that escaped during two separate swings was the same, he knew that both had taken an equal time.

His experiments showed that the time of swing depends only on the length of the pendulum. To double the time of swing you must make your pendulum four times as long; to treble the time of swing you must make your pendulum nine times as long. The length of the pendulum varies in the same ratio as the square of the time of swing. (We now know that this rule, while correct for narrow swings, is not quite accurate when the pendulum swings through a very wide arc.) In 1657 the Dutch scientist Christian Huygens used Galileo's discovery to produce accurate pendulum clocks.

Before Galileo's time, people believed that the heavier an object was, the faster it would fall. Galileo's pendulum experiments, however, disproved this. For he found that the weight of the bob at the bottom of the pendulum has no effect on the time of swing. To settle the matter beyond dispute, he dropped two different weights simultaneously from the Leaning Tower of Pisa. The heavy one and the light one hit the ground at the same instant.

151

The Rhythm of the Pendulum

by Galileo Galilei

In this book, Dialogues Concerning Two New Sciences, *Galileo uses Salviati as his mouthpiece to explain the law of the pendulum in a conversation with Sagredo.*

Salv. As to the times of vibration of bodies suspended by threads of different lengths, they bear to each other the same proportion as the square roots of the lengths of the thread; or one might say the lengths are to each other as the squares of the times; so that if one wishes to make the vibration-time of one pendulum twice that of another, he must make its suspension four times as long. In like manner, if one pendulum has a suspension nine times as long as another, this second pendulum will execute three vibrations during each one of the first; from which it follows that the lengths of the suspending cords bear to each other the [inverse] ratio of the squares of the number of vibrations performed in the same time.

      Sagr. Then, if I understand you correctly, I can easily measure the length of a string whose upper end is attached at any height whatever even if this end were invisible and I could see only the lower extremity. For if I attach to the lower end of this string a rather heavy weight and give it a to-and-fro motion, and if I ask a friend to count a number of its vibrations, while I, during the same time-interval, count the number of vibrations of a pendulum which is exactly one cubit in length, then knowing the number of vibrations which each pendulum makes in the given interval of time one can determine the length of the string. Suppose, for example, that my friend counts 20 vibrations of the long cord during the same time in which I count 240 of my string which is one cubit in length; taking the squares of the two numbers, 20 and 240, namely 400 and 57600, then, I say, the long string contains 57600 units of such length that my pendulum will contain 400 of them; and since the length of my string is one·cubit, I shall divide 57600 by 400 and thus obtain 144. Accordingly I shall call the length of the string 144 cubits.

Salv. Nor will you miss it by as much as a hand's breadth, especially if you observe a large number of vibrations.

Sagr. You give me frequent occasion to admire the wealth and profusion of nature when, from such common and even trivial phenomena, you derive facts which are not only striking and new but which are often far removed from what we would have imagined. Thousands of times I have observed vibrations especially in churches where lamps, suspended by long cords, had been inadvertently set into motion; but the most which I could infer from these observations was that the view of those who think that such vibrations are maintained by the medium is highly improbable: for, in that case, the air must needs have considerable judgment and little else to do but kill time by pushing to and fro a pendent weight with perfect regularity. But I never dreamed of learning that one and the same body, when suspended from a string a hundred cubits long and pulled aside through an arc of 90° or even 1° or $\frac{1}{2}$°, would employ the same time in passing through the least as through the largest of these arcs.

The Rhythms of Rational Numbers

by Irving Adler

Numbers, too, have their rhythms. In this selection from
A New Look at Arithmetic, *Irving Adler explains the significance
of the rhythms that appears in some non-terminating decimals like*
.33333 . . . , .14141414 . . . , *and so on.*

FROM DECIMAL FRACTION TO COMMON FRACTION

Every decimal fraction can be written as a single common fraction
whose denominator is a power of ten. If the decimal fraction has
one decimal place, the denominator is 10^1; if the decimal fraction
has two decimal places, the denominator is 10^2; in general, if the
decimal fraction has n decimal places, the denominator is 10^n. For
example,

$$.2 = \frac{2}{10} \; ; \; .25 = \frac{25}{10^2} = \frac{25}{100} \; ; \; .275 = \frac{275}{10^3} = \frac{275}{1000} \text{ , etc.}$$

The common fraction obtained in this way may not be in lowest
terms. An equivalent fraction that is in lowest terms may be obtained
in the usual way by dividing the numerator and the denominator by
their greatest common divisor. Thus

$$\frac{2}{10} = \frac{2 \div 2}{10 \div 2} = \frac{1}{5} \; ; \; \frac{25}{100} = \frac{25 \div 25}{100 \div 25} = \frac{1}{4} \text{ ; etc.}$$

FROM COMMON FRACTION TO DECIMAL FRACTION

While every decimal fraction can be converted into a common
fraction, the converse is not true. A common fraction can be converted into a decimal fraction only if its denominator is a factor of
some power of 10.

154

Example: The denominator of the fraction $\frac{3}{5}$ is a factor of 10. So $\frac{1}{3}$ can be changed into an equivalent fraction whose denominator is 10:

$$\frac{3}{5} = \frac{3 \times 2}{5 \times 2} = \frac{6}{10} = .6.$$

Example: The denominator of the fraction $\frac{2}{25}$ is a factor of 100. So $\frac{2}{25}$ can be changed into an equivalent fraction whose denominator is

$$100: \frac{2}{25} = \frac{2 \times 4}{25 \times 4} = \frac{8}{100} = .08.$$

Example: The denominator of the fraction $\frac{1}{3}$ is not a factor of any power of 10. So there is no decimal fraction that is equivalent to $\frac{3}{5}$.

In the examples above, we converted $\frac{3}{5}$ and $\frac{2}{25}$ into decimal fractions, by first changing them to equivalent common fractions whose denominators are powers of 10, and then shifting to the decimal-fraction notation. There is another way of making the conversion that is based on the fact that the fraction line may be interpreted to be a division sign. Then $\frac{3}{5}$ means $3 \div 5$, and $\frac{2}{25}$ means $2 \div 25$. To find the decimal fraction that is equivalent to each of these common fractions, simply carry out the indicated division, keeping in mind that 3 is the same as 3.0 or 3.00 or 3.000, etc., and 2 is the same as 2.0, or 2.00, or 2.000, etc. Thus,

$$\begin{array}{r} .6 \\ 5\overline{)3.0} \\ \underline{3.0} \end{array} \qquad \text{So } \frac{3}{5} = .6$$

$$\begin{array}{r} .08 \\ 25\overline{)2.00} \\ \underline{00} \\ 200 \\ \underline{200} \end{array} \qquad \text{So } \frac{2}{25} = .08$$

NON-TERMINATING DECIMALS

The second method of converting a common fraction into a decimal fraction suggests that we make another try with a fraction like $\frac{1}{3}$. Just as we may interpret $\frac{3}{5}$ to mean $3 \div 5$, we may also interpret $\frac{1}{3}$ to mean $1 \div 3$. Then, by writing 3 as 3.0, 3.00, 3.000, etc., we may carry out the division to one, two, three decimal places, and so on. In the division exercise $3 \div 5$, the process terminates, and we end up with the decimal fraction .6. A decimal fraction with a finite number of decimal places is sometimes referred to as a terminating decimal. In the division exercise $1 \div 3$, the process does not terminate, because after each subtraction step in the division algorithm there is a remainder, so the division can be carried out one more step to obtain another decimal place in the quotient. If we imagine the division algorithm carried out indefinitely, we obtain what looks like a decimal fraction with an infinite number of decimal places, and is called a non-terminating decimal. In this case the non-terminating decimal is .3333 ... , where the three dots indicate that the decimal does not terminate. There is a strong temptation to say that just as the terminating decimal .6 represents the fraction $\frac{3}{5}$, the non-terminating decimal .3333 ... represents the fraction $\frac{1}{3}$. We may yield to this temptation only if we can give some definite meaning to the statement that .3333 ... represents the fraction $\frac{1}{3}$.

We give meaning to the statement by first interpreting the non-terminating decimal as simply an abbreviated way of writing the sequence of terminating decimals obtained by using first one decimal place, then two decimal places, then three decimal places, etc. Thus .3333 ... is understood to be an abbreviation for the sequence .3, .33, .333, .3333, .33333, ... where the three dots here indicate that the sequence does not terminate. It is not difficult to relate this sequence to the fraction $\frac{1}{3}$. The first term of the sequence, .3, is an approximation to $\frac{1}{3}$. The difference between $\frac{1}{3}$ and this approximation is $\frac{1}{3} - \frac{3}{10} = \frac{1}{30}$. The second term in the sequence, .33, is a better approximation because $\frac{1}{3} - \frac{33}{100} = \frac{1}{300}$. The third term of the sequence, .333, is a still better approximation, because $\frac{1}{3} - \frac{333}{1000} = \frac{1}{3000}$; and so on. The successive terms of the sequence of terminating decimals .3, .33, .333, ... give better and better approximations of the fraction $\frac{1}{3}$, with the error shrinking towards zero as we move along from term to term in the sequence. We may therefore say that the non-terminating decimal .333 ... "represents" the fraction $\frac{1}{3}$ in this sense: the terminating decimals obtained by using only the first decimal place, then only the first two decimal places, then only the first three decimal places, etc., give better and better approximations of $\frac{1}{3}$, with the error shrinking toward zero as more and more decimal places are used.

Any common fraction that is not equivalent to a terminating decimal may be represented by a non-terminating decimal. The non-terminating decimal is obtained by simply dividing the numerator by the denominator by the usual long-division algorithm.

156

Example: Represent $\frac{211}{990}$ by a non-terminating decimal.

$$
\begin{array}{r}
.21313 \\
990\overline{)211.00000} \\
\underline{198\ 0} \\
13\ 00 \\
\underline{9\ 90} \\
3\ 100 \\
\underline{2\ 970} \\
1300 \\
\underline{990} \\
3100 \\
\underline{2970} \\
130
\end{array}
$$

Notice that, in the third subtraction step in the algorithm, the remainder is 130, the same as the remainder that resulted from the first subtraction step. Consequently, from this point on the algorithm begins to repeat the steps that occur between the first subtraction and the third subtraction. As a result, the sequence of digits 13 which occurs in the second and third decimal places is repeated indefinitely. The fraction $\frac{211}{990}$ is therefore represented by the non-terminating decimal .2131313 . . . , where the three dots indicate indefinite repetition of the pair of digits 13. A non-terminating decimal which, after a finite number of decimal places, begins to repeat a particular sequence of digits over and over again is called a *repeating decimal.* To indicate a repeating decimal with a minimum of writing, it is customary to write only enough decimal places to include the repeating part once, and to identify the repeating part by underlining it. Thus the repeating decimal for $\frac{211}{990}$ is written as .21<u>3</u> . . . .Whenever a fraction is converted into a non-terminating decimal, the decimal is a repeating decimal. The proof of this assertion is given in the book *The New Mathematics.**

When a fraction is represented by a terminating decimal, we may also represent it by a non-terminating repeating decimal by simply appending more decimal places filled with a string of zeros. For example, $\frac{3}{5} = .6 = .6000 . . . = .6\underline{0} . . . .$ We can also convert a terminating decimal into a non-terminating repeating decimal in another way: Reduce the digit in the last decimal place by 1, and then append more decimal places filled with a string of nines. For example, $\frac{3}{5} = .6 = .5999 . . . = .5\underline{9} . . . .$ Consequently, every fraction,

The New Mathematics by Irving Adler. 1958. John Day. Also in paperback. 1960. Sig. NAL.

without exception, can be represented by a non-terminating repeating decimal. This conclusion applies equally to proper and improper fractions and to positive and negative fractions. Thus, $\frac{8}{5} = 1\frac{3}{5} = 1.6 = 1.5\underline{9}\ldots$, and $-\frac{8}{5} = -1.5\underline{9}\ldots$. So we may say that *every rational number can be represented by a non-terminating repeating decimal.*

FROM REPEATING DECIMAL TO COMMON FRACTION

We have seen that every common fraction can be converted into a repeating decimal. Can every repeating decimal be converted into a common fraction? The answer is *yes*, and there is a simple way to do it. The method is best explained by giving some examples of its use.

Example: Convert $3.5\underline{6}\ldots$ into a common fraction. $3.5\underline{6}\ldots$ means $3.56666\ldots$ where the digit 6 is repeated indefinitely.

Let $x = 3.56666\ldots$. Multiplying by 10, we get

$$10x = 35.6666\ldots$$
$$x = 3.5666\ldots$$

Subtract the second equation from the first. We get $9x = 32.1$, a finite decimal, because the decimals $35.6666\ldots$ and $3.5666\ldots$ agree in all decimal places after the first decimal place. To convert 32.1 into a whole number, multiply both sides of the equation by 10. Then we have $90x = 321$, and therefore $x = \frac{321}{90}$. The infinite decimal $3.5666\ldots$ represents the common fraction $\frac{321}{90}$. This can be verified by dividing 321 by 90.

Example: Convert $2.1\underline{7}\ldots$ into a common fraction. $2.1\underline{7}\ldots$ means $2.171717\ldots$ where the sequence 17 is repeated indefinitely. Let $x = 2.171717\ldots$. This time, since there are two digits in the repeating part of the decimal, we multiply by 10^2 or 100. (If the repeating part contains three digits, multiply by 10^3 or 1000. If the repeating part contains n digits, multiply by 10^n.) Then we have

$$100x = 217.1717\ldots$$
$$x = 2.1717\ldots$$

Subtracting, we get $99x = 215$, and $x = \frac{215}{99}$. Since this procedure can obviously be followed with every repeating decimal, we have the important conclusion that every repeating decimal represents a rational number.

158

Exercises:
1. Use the law of one to change $\frac{4}{125}$ to a decimal fraction.
2. Use long-division to change $\frac{3}{125}$ to a decimal fraction.
3. Find a non-terminating repeating decimal that represents:
   (a) $\frac{4}{9}$ (b) $\frac{2}{15}$ (c) $\frac{3}{7}$
4. Convert into a common fraction:
   (a) .7̲7 ...   (b) .2̲5̲5 ...
   (c) .1̲6̲16 ...   (d) .41̲2̲12 .. ⸴

Answers

1. $\dfrac{4}{125} = \dfrac{4}{125} \times 1 = \dfrac{4}{125} \times \dfrac{8}{8} = \dfrac{32}{1000} = .032$

2. $\quad .024$
   $125\overline{)3.000}$
   $\qquad \underline{250}$
   $\qquad \ \ 500$
   $\qquad \ \ \underline{500}$

3. (a) .4̅ ···   (b) .1̅3̅ ···   (c) .4̅2̅8̅5̅7̅1̅ ···

4. (a) $\dfrac{6}{7}$; (b) $\dfrac{6}{23}$; (c) $\dfrac{16}{99}$; (d) $\dfrac{408}{990}$

PART 10

Discovery

Making Your Own Discoveries

You don't have to be a mathematician to make mathematical discoveries. You, too, can make your own discoveries by following the simple procedures usually followed by mathematicians themselves.

Here are the usual procedures: *Step 1:* Do something. *Step 2:* Observe the results of what you have done. *Step 3:* Look for a pattern in the results. Try to guess what the pattern is. *Step 4:* Test your guess or hunch to see if it seems reasonable. *Step 5:* Prove that your guess is right. Step 5 may be difficult to carry out. Even if you don't succeed in finishing step 5, you can have the satisfaction of making a meaningful mathematical discovery.

To demonstrate the procedure, let us try to answer the question, "What kind of sum do you get when you add any two odd numbers?"

Step 1. Add some odd numbers. For example, $3 + 5 = 8$; $7 + 13 = 20$; $21 + 19 = 40$; $9 + 45 = 54$.

Step 2: The results we got were 8, 20, 40, 54.

Step 3: It looks as though the sum of two odd numbers is always an even number.

Step 4 Let's check out our guess with a few more cases: $1 + 3 = 4$; $5 + 7 = 12$; $3 + 13 = 16$. So far, so good.

Step 5: Every even number is double some whole number, so it may be represented by $2a$, where a is some whole number, or $2b$, or $2c$, etc. Every odd number is one more than an even number, so it may be represented by $2a + 1$, or $2b + 1$, etc. If $2a + 1$ is one odd number and $2b + 1$ is another odd number, then their sum is $(2a + 1) + (2b + 1) = 2a + 2b + 1 + 1 = 2a + 2b + 2 \times 1 = 2 (a + b + 1)$. That is, their sum is double a whole number, so it must be an even number.

162

Here are three problems that give you an opportunity to make your own discoveries. In order not to spoil your fun, we shall not give you the solutions to these problems. If you want the solutions, you'll have to find them yourself.

Problem 1. This problem concerns four-digit numbers. We shall include among them all numbers with fewer than four digits by putting enough zeros in on the left to turn them into four-digit numbers. For example, the number 12 can be written as the four-digit number 0012. Now here is the problem: take any four-digit number except one in which all four digits are the same, that is, except 0000, 1111, 2222, etc. Arrange the digits of this number in descending order and in ascending order to form two new numbers, and then subtract the smaller of these two from the larger. Using the result of this procedure, start all over again, and *repeat the procedure over and over again, and see what happens.* For example, if you start with 2598, you write 9852 − 2598 = 7263. Now repeat the procedure with 7263: 7632 − 2367 = 5265. Then repeat the procedure with 5265, and so on.

Problem 2. An endless sequence of digits in which each digit is either 0 or 1 is called a *binary sequence.* If the sequence consists of a group of digits repeated over and over again, it is called a *periodic* sequence, and the number of digits in the repeated group is called the *period* of the sequence. For example, the binary sequence,

$$10111011101110111011\ldots$$

is periodic because the group of digits 1011 is repeated over and over again to form it. The period of this sequence is four. The complement of a binary sequence is another sequence obtained by replacing each 1 by a 0, and each 0 by a 1. Thus, the complement of 10111011 ... is 01000100 ... Now here is the problem: Take any periodic binary sequence and replace it by a new sequence obtained in this way: Replace a digit by 0 if the digit is the same as the digit that follows it in the sequence; replace a digit by 1 if the digit is not the same as the digit that follows it in the sequence. Repeat this procedure with the new sequence. *Repeat the procedure over and over again, and see what happens.* For example, if you start with the sequence 1010101010 ... you get these sequences in succession: 11111 ... , 00000 ... , 00000 ... , etc.

Problem 3. Form a new number from any whole number by adding the cubes of its digits. *Repeat the procedure over and over again, and see what happens.* For example, if you start with the number 13, you write:

$$1^3 + 3^3 = 1 + 27 = 28; \quad 2^3 + 8^3 = 8 + 512 = 520;$$
$$5^3 + 2^3 + 0^3 = 125 + 8 + 0 = 133;$$
$$1^3 + 3^3 + 3^3 = 1 + 27 + 27 = 55; \text{ etc.}$$

Srinivasa Ramanujan
(1887-1920)

Woodcarving by William Ransom

"... [Ramanujan] was not particularly interested in his own history or psychology; he was a mathematician anxious to get on with the job. And after all I too was a mathematician, and a mathematician meeting Ramanujan had more interesting things to think about than historical research. It seemed ridiculous to worry him about how he had found this or that known theorem, when he was showing me half a dozen new ones almost every day."[1]

[1] G. H. Hardy, *Ramanujan,* Cambridge University Press, 1940, p. 11.

Srinivasa Ramanujan, whose picture is reproduced on this page, was a self-taught Indian mathematician with great creative ability. Working alone, without any formal training, and without the benefit of an exchange of ideas with other mathematicians, he made many important and interesting discoveries. The fascinating story of Ramanujan is told in the next selection, by James Newman. It first appeared in Scientific American *in 1948.*

Ramanujan and His Discoveries*

by James R. Newman

SRINIVASA RAMANUJAN AIYANGAR, according to his biographer Seshu Aiyar, was a member of a Brahman family in somewhat poor circumstances in the Tanjore district of the Madras presidency. His father was an accountant to a cloth merchant at Kumbakonam, while his mother, a woman of "strong common sense," was the daughter of a Brahman petty official in the Munsiff's (or legal judge's) Court at Erode. For some time after her marriage she had no children, "but her father prayed to the famous goddess Namagiri, in the neighborhood town of Namakkal, to bless his daughter with offspring. Shortly afterwards, her eldest child, the mathematician Ramanujan, was born on 22nd December 1887."

He first went to school at five and was transferred before he was seven to the Town High School at Kumbakonam, where he held a scholarship. His extraordinary powers appear to have been recognized almost immediately. He was quiet and meditative and had an extraordinary memory. He delighted in entertaining his friends with theorems and formulae, with the recitation of complete lists of Sanskrit roots and with repeating the values of *pi* and the square root of two to any number of decimal places.

When he was 15 and in the sixth form at school, a friend of his secured for him the loan of Carr's *Synopsis of Pure Mathematics* from the library of the local Government College. Through the new world thus opened to him Ramanujan ranged with delight. It was this book that awakened his genius. He set himself at once to establishing its formulae. As he was without the aid of other books, each solution was for him a piece of original research. He first devised methods for constructing magic squares. Then he branched off to geometry, where he took up the squaring of the circle and went so far as to get a result for the length of the equatorial circumference of the earth which differed from the true length by only a few feet. Finding the scope of geometry limited, he turned his attention to algebra. Ramanujan used to say that the goddess of Namakkal inspired him with the formulae in dreams. It is a remarkable fact that, on rising from bed, he would frequently note down results and verify them, though he was not always able to supply a rigorous proof. This pattern repeated itself throughout his life.

He passed his matriculation examination to the Government College at Kumbakonam at 16, and secured the "Junior Subrahmanyam Scholarship." Owing to weakness in English — for he gave no thought to anything but mathematics — he failed in his next examination and lost his scholarship. He then left Kumbakonam, first for Vizagapatam and then for Madras. Here he presented himself for the "First Examination in Arts" in December 1906, but failed and never tried again. For the next few years he continued his independent work in mathematics. In 1909 he was married and it became necessary for him to find some permanent employment. In the course of his search for work he was given a letter of recommendation to a true lover of mathematics, Diwan Bahadur R. Ramanchandra Rao, who was then Collector at Nelore, a small town 80 miles north of Madras. Ramachandra Rao had already seen one of the two fat notebooks kept by Ramanujan into which he crammed his wonderful ideas. His first interview with Ramanujan is best described in his own words.

"Several years ago, a nephew of mine perfectly innocent of mathematical knowledge said to me, 'Uncle, I have a visitor who talks of mathematics; I do not understand him; can you see if there is anything in his talk?' And in the plenitude of my mathematical wisdom, I condescended to permit Ramanujan to walk into my presence. A short uncouth figure, stout, unshaved, not overclean, with one conspicuous feature — shining eyes — walked in with a frayed notebook under his arm. He was miserably poor. He had run away from Kumbakonam to get leisure in Madras to pursue his studies. He never craved for any distinction. He wanted leisure; in other words, that simple food should be provided for him without exertion on his part and that he should be allowed to dream on.

"He opened his book and began to explain some of his

discoveries. I saw quite at once that there was something out of the way; but my knowledge did not permit me to judge whether he talked sense or nonsense. Suspending judgment.·I asked him to come over again, and he did. And then he had gauged my ignorance and showed me some of his simpler results. These transcended existing books and I had no doubt that he was a remarkable man. Then, step by step, he led me to elliptic integrals and hypergeometric series and at last his theory of divergent series not yet announced to the world converted me. I asked him what he wanted. He said he wanted a pittance to live on so that he might pursue his researches."

Ramachandra Rao undertook to pay Ramanujan's expenses for a time. After a while, other attempts to obtain a scholarship having failed and Ramanujan being unwilling to be supported by anyone for any length of time, he accepted a small appointment in the office of the Madras Port Trust.

But he never slackened his work in mathematics. His earliest contribution was published in the *Journal of the Indian Mathematical Society* in 1911, when Ramanujan was 23. His first long article was on "Some Properties of Bernoulli's Numbers" and was published in the same year. In 1912 he contributed two more notes to the same journal and also several questions for solution.

By this time Ramachandra Rao had induced a Mr. Griffith of the Madras Engineering College to take an interest in Ramanujan, and Griffith spoke to Sir Francis Spring, the chairman of the Madras Port Trust, where Ramanujan was employed. From that time on it became easy to secure recognition of his work. Upon the suggestion of Seshu Aiyar and others, Ramanujan began a correspondence with G. H. Hardy, then Fellow of Trinity College, Cambridge. His first letter to Hardy, dated January 16, 1913, which his friends helped him put in English, follows:

"Dear Sir,

"I beg to introduce myself to you as a clerk in the Accounts Department of the Port Trust Office at Madras on a salary of only £20 per annum. I am now about 23 years of age. [*He was actually 25 — Ed.*] I have had no University education but I have undergone the ordinary school course. After leaving school I have been employing the spare time at my disposal to work at Mathematics. I have not trodden through the conventional regular course which is followed in a University course, but I am striking out a new path for myself. I have made a special investigation of divergent series in general and the results I get are termed by the local mathematicians as 'startling'. . . .

"I would request you to go through the enclosed papers. Being poor, if you are convinced that there is anything of value I would like to have my theorems published. I have not given the actual investigations nor the expressions that I get but I have indicated the lines on which I proceed. Being inexperienced I would very highly value any advice you give me. Requesting to be excused for the trouble I give you.

"I remain, Dear Sir, Yours truly,

"S. Ramanujan."

To the letter were attached about 120 theorems, about which Hardy later made these comments: "A single look at them is enough to show that they could only be written down by a mathematician of the highest class. They must be true because, if they were not true, no one would have had the imagination to invent them. Finally . . . the writer must be completely honest, because great mathematicians are commoner than thieves or humbugs of such incredible skill. . . ."

"While Ramanujan had numerous brilliant successes, his work on prime numbers and on all the allied problems of the theory was definitely wrong. This may be said to have been his one great failure. And yet I am not sure that, in some ways, his failure was not more wonderful than any of his triumphs. . . ."

Ramanujan's notation of one mathematical term in this area, wrote Hardy, "was first obtained by Landau in 1908. Ramanujan had none of Landau's weapons at his command; he had never seen a French or German book; his knowledge even of English was insufficient to qualify for a degree. It is sufficiently marvellous that he should have even dreamt of problems such as these, problems which it had taken the finest mathematicians in Europe a hundred years to solve, and of which the solution is incomplete to the present day."

At last, in May of 1913, as the result of the help of many friends, Ramanujan was relieved of his clerical post in the Madras Port Trust and given a special scholarship. Hardy had made efforts from the first to bring Ramanujan to Cambridge. The way seemed to be open, but Ramanujan refused at first to go because of caste prejudice and lack of his mother's consent.

"This consent," wrote Hardy, "was at last got very easily in an unexpected manner. For one morning his mother announced that she had had a dream on the previous night, in which she saw her son seated in a big hall amidst a group of Europeans, and that the goddess Namagiri had commanded her not to stand in the way of her son fulfilling his life's purpose."

When Ramanujan finally came, he had a scholarship from Madras of £250, of which £50 was allotted to the support of his family in India, and an allowance of £60 from Trinity.

"There was one great puzzle," Hardy observes of Ramanujan. "What was to be done in the way of teaching him modern mathematics? The limitations of his knowledge were as startling as its profundity. Here was a man who could work out modular equations, and theorems of complex multiplication, to orders unheard of, whose mastery of continued fractions was, on the formal side at any rate, beyond that of any mathematician in the world, who had found for himself the functional equation of the Zeta-function and the dominant terms of many of the most famous problems in the analytic theory of numbers; and he had never heard of a doubly periodic function or of Cauchy's theorem, and had indeed but the vaguest idea of what a function of a complex variable was. His ideas as to what constituted a mathematical proof were of the most shadowy description. All his results, new or old, right or wrong, had been arrived at by a process of mingled argument, intuition, and induction, of which he was entirely unable to give any coherent account.

"It was impossible to ask such a man to submit to systematic instruction, to try to learn mathematics from the beginning once more. I was afraid too that, if I insisted unduly on matters which Ramanujan found irksome, I might destroy his confidence or break the spell of his inspiration. On the other hand there were things of which it was impossible that he should remain in ignorance. Some of his results were wrong, and in particular those which concerned the distribution of primes, to which he attached the greatest importance. It was impossible to allow him to go through life supposing that all the zeros of the Zeta-function were real. So I had to try to teach him, and in a measure I succeeded, though obviously I learnt from him much more than he learnt from me. . . .

"I should add a word here about Ramanujan's interest outside mathematics. Like his mathematics, they shewed the strangest contrasts. He had very little interest, I should say, in literature as such, or in art, though he could tell good literature from bad. On the other hand, he was a keen philosopher, of what

169

appeared, to followers of the modern Cambridge school, a rather nebulous kind, and an ardent politician, of a pacifist and ultra-radical type. He adhered, with a severity most unusual in Indians resident in England, to the religious observances of his caste; but his religion was a matter of observance and not of intellectual conviction, and I remember well his telling me (much to my surprise) that all religions seemed to him more or less equally true. Alike in literature, philosophy, and mathematics, he had a passion for what was unexpected, strange, and odd; he had quite a small library of books by circle-squarers and other cranks... He was a vegetarian in the strictest sense — this proved a terrible difficulty later when he fell ill — and all the time he was in Cambridge he cooked all his food himself, and never cooked it without first changing into pyjamas. ...

"It was in the spring of 1917 that Ramanujan first appeared to be unwell. He went to a Nursing Home at Cambridge in the early summer, and was never out of bed for any length of time again. He was in sanatoria at Wells, at Matlock, and in London, and it was not until the autumn of 1918 that he shewed any decided symptom of improvement. He had then resumed active work, stimulated perhaps by his election to the Royal Society, and some of his most beautiful theorems were discovered about this time. His election to a Trinity Fellowship was a further encouragement; and each of those famous societies may well congratulate themselves that they recognized his claims before it was too late."

Early in 1919 Ramanujan went home to India, where he died in the following year.

For an evaluation of Ramanujan's methods and work in mathematics we must again quote from Hardy. "I have often been asked whether Ramanujan had any special secret; whether his methods differed in kind from those of other mathematicians; whether there was anything really abnormal in his mode of thought. I cannot answer these questions with any confidence or conviction; but I do not believe it. My belief is that all mathematicians think, at bottom, in the same kind of way, and that Ramanujan was no exception. He had, of course, an extraordinary memory. He could remember the idiosyncrasies of numbers in an almost uncanny way. It was Mr. Littlewood (I believe) who remarked that 'every positive integer was one of his personal friends.' I remember once going to see him when he was lying ill at Putney. I had ridden in taxi-cab No. 1729, and remarked that

170

the number seemed to me rather a dull one, and that I hoped it was not an unfavourable omen. 'No,' he replied, 'it is a very interesting number; it is the smallest number expressible·as a sum of two cubes in two different ways.' I asked him, naturally, whether he knew the answer to the corresponding problem for fourth powers; and he replied, after a moment's thought, that he could see no obvious example, and thought that the first such number must be very large. His memory, and his powers of calculation, were very unusual, but they could not reasonably be called 'abnormal.' If he had to multiply two large numbers, he multiplied them in the ordinary way; he could do it with unusual rapidity and accuracy, but not more rapidly or more accurately than any mathematician who is naturally quick and has the habit of computation.

"It was his insight into algebraical formulae, transformations of infinite series, and so forth, that was most amazing. On this side most certainly I have never met his equal, and I can compare him only with Euler or Jacobi. He worked, far more than the majority of modern mathematicians, by induction from numerical examples; all of his congruence properties of partitions, for example, were discovered in this way. But with his memory, his patience, and his power of calculation, he combined a power of generalisation, a feeling for form, and a capacity for rapid modification of his hypotheses, that were often really startling, and made him, in his own field, without a rival in his day.

"It is often said that it is much more difficult now for a mathematician to be original than it was in the great days when the foundations of modern analysis were laid; and no doubt in a measure it is true. Opinions may differ as to the importance of Ramanujan's work, the kind of standard by which it should be judged, and the influence which it is likely to have on the mathematics of the future. It has not the simplicity and the inevitableness of the very greatest work; it would be greater if it were less strange. One gift it has which no one can deny — profound and invincible originality. He would probably have been a greater mathematician if he had been caught and tamed a little in his youth; he would have discovered more that was new, and that, no doubt, of greater importance. On the other hand he would have been less of a Ramanujan, and more of a European professor and the loss might have been greater than the gain."

Indeterminate Problems

by Oystein Ore

There is a certain kind of puzzle that has come up over and over again through the centuries in many different parts of the world. People usually discover the solution to this kind of puzzle by trial and error. But these puzzles can also be solved in a systematic way by a method developed by the great mathematician Leonard Euler (1707-1783). Many of these puzzles, and Euler's method of solving them, are given in the next selection, taken from Oystein Ore's Number Theory and Its History.

There is a type of problem that occurs quite commonly in puzzles and whose theory constitutes a particularly significant part of number theory. These problems may appropriately be called *linear indeterminate problems*, for reasons that will become clear after some examples.

One of the earliest occurrences of such problems in Europe is to be found in a manuscript containing mathematical problems dating from about the tenth century. It is believed possible that it may be a copy of a collection of puzzles which Alcuin prepared for Charlemagne. The problem we are interested in runs as follows:

1. When 100 bushels of grain are distributed among 100 persons so that each man receives three bushels, each woman two bushels, and each child half a bushel, how many men, women, and children are there?

To formulate this problem mathematically let x, y, and z denote the number of men, women, and children, respectively. The conditions of the problem then give

$$x + y + z = 100, \qquad 3x + 2y + \tfrac{1}{2}z = 100 \qquad (6\text{-}1)$$

172

As we shall see later there are several solutions but Alcuin gives only the values

$$x = 11, \qquad y = 15, \qquad z = 74$$

From an Arabic manuscript copied about A.D. 1200, but undoubtedly composed earlier, we take this example:

2. One duck may be bought for 5 drachmas, one chicken for 1 drachma, and 20 starlings for 1 drachma. You are given 100 drachmas and ordered to buy 100 birds. How many will there be of each kind?

When x, y, and z are the number of ducks, chickens, and starlings, it follows that

$$x + y + z = 100, \qquad 5x + y + \frac{z}{20} = 100 \qquad (6\text{-}2)$$

One may observe that the same number occurs on the right-hand side in both equations (6-1) and in (6-2). This particular preference in the choice of the figures in the questions is common in Arabic, Chinese, and medieval European problems, and it undoubtedly points to an interrelated or common background. Even the use of the special number 100 shows a peculiar persistence in problems from all these sources.

One finds similar questions in the many medieval collections of problems. They occur in Leonardo's *Liber Abaci* (A.D. 1202), probably derived from Arabic sources, and in the following centuries they became increasingly popular. To illustrate a fairly common type of formulation we quote from a German reckoning manual (Christoff Rudolff, 1526):

3. At an inn, a party of 20 persons pay a bill of 20 groschen. The party consists of men (x), women (y), and maidens (z), each man paying 3, each woman 2, and each maiden $\frac{1}{2}$ groschen. How was the party composed?

Here the equations become

$$x + y + z = 20, \qquad 3x + 2y + \frac{z}{2} = 20 \qquad (6\text{-}3)$$

and the figures are so chosen that there is a unique solution $x = 1$, $y = 5, z = 14$.

It is, of course, not certain that this type of problem originated within a single cultural sphere, but if so, it seems likely that India should be looked to for its source. As early as the arithmetic of Aryabhata (around A.D. 500) one finds indeterminate problems. Brahmagupta (born A.D. 598) in his mathematical and astronomical

173

manual *Brahma-Sphuta-Siddhanta* ("Brahma's correct system") not only introduces them, but gives a perfected method for their solution that is practically equivalent to our present procedures. The method is called the *cuttaca* or *pulverizer* and is based upon Euclid's algorism. Brahmagupta's examples are almost all of astronomical character and refer to the comparisons between periods of revolution of the heavenly bodies and determinations of their relative positions.

We take the following problem from the *Lilavati* by Bhaskara, a work we have already mentioned:

4. Say quickly, mathematician, what is the multiplier by which two hundred and twenty-one being multiplied and sixty-five added to the product the sum divided by one hundred and ninety-five becomes exhausted?

Here one wishes to find some x satisfying the condition

$$221x + 65 = 195y \qquad (6\text{-}4)$$

In the *Lilavati*, as well as in other Hindu treatises on mathematics, one finds many problems in the flowery style so customary in Hindu writings. This problem is from the *Bija-Ganita*, literally meaning *seed-counting* but denoting *algebra*, also composed by Bhaskara:

5. The quantity of rubies without flaw, sapphires, and pearls belonging to one person is five, eight, and seven respectively; the number of like gems appertaining to another is seven, nine and six; in addition, one has ninety-two coins, the other sixty-two and they are equally rich. Tell me quickly then, intelligent friend, who art conversant with algebra, the prices of each sort of gem.

In Hindu mathematics colors were used to denote the various unknowns, black, blue, yellow, red, and so on. If we prosaically denote the prices of rubies, sapphires, and pearls by x, y, and z, the condition becomes

$$5x + 8y + 7z + 92 = 7x + 9y + 6z + 62 \qquad (6\text{-}5)$$

A further example may be taken from Mahaviracarya's work *Ganita-Sara-Sangraha*, probably composed around A.D. 850:

6. Into the bright and refreshing outskirts of a forest which were full of numerous trees with their branches bent down with the weight of flowers and fruits, trees such as jambu trees, date-palms, hintala trees, palmyras, punnaga trees and mango trees — filled with the many sounds of crowds of parrots and cuckoos found

near springs containing lotuses with bees roaming around them — a number of travelers entered with joy.

There were 63 equal heaps of plantain fruits put together and seven single fruits. These were divided evenly among 23 travelers. Tell me now the number of fruits in each heap.

It is quite an anticlimax to state that if x is the number of fruits in each heap, one must have

$$63x + 7 = 23y \qquad (6\text{-}6)$$

This beautiful Hindu forest contains a number of other problems, but after all these ancient examples let us conclude with one with a more modern touch. The following letter was one among several similar ones received by the author during the recent war:

Dear Sir:

A group of bewildered GI's at Guadalcanal, most of whom have been out of school for a good many years and have forgotten how to solve algebraic problems, have been baffled by what appears to be a very simple problem. Some of them affirm that it cannot be worked other than through the trial and error method, but I maintain that it can be worked systematically by means of some sort of formula or equation.

[7.] Here is the problem: A man has a theater with a seating capacity of 100. He wishes to admit 100 people in such a proportion that will enable him to take in $1.00 with prices as follows: men 5¢, women 2¢, children 10 for one cent. How many of each must be admitted?

Can this problem be solved other than through the laborious trial and error method? We shall greatly appreciate your assistance in helping us to find the solution, thus relieving our weary brains.

Yours truly,

P.S. Through the trial and error method we found the answer to be 11 men, 19 women, and 70 children.

In terms of equations we have the conditions

$$x + y + z = 100, \qquad 5x + 2y + \frac{z}{10} = 100 \qquad (6\text{-}7)$$

Here again our familiar number 100 figures on the right. It is also clear that the figures are not adapted to the present movie prices

175

and that we are confronted with an ancient problem which has gained in actuality by being put in modern dress.

Problems with two unknowns. We have presented a whole series of examples of linear indeterminate problems. As we saw, they lead to one or more linear equations between the unknown quantities. Furthermore, the number of unknowns is greater than the number of equations so that if there were no limitations on the kind of values the solutions could take, one could give arbitrary values to some of the unknowns and find the others in terms of them. For instance, in problem 4 one could simply write

$$x = \frac{195y - 65}{221}$$

and any value of y would give a corresponding value of x. However, by the terms of the problems, the choice of solutions is limited to integral values and usually also to positive numbers. But even with these restrictions, the solutions may be indeterminate in the sense that there may be several, or even an infinite number of them, as we shall see. On the other hand there may be no solution at all.

Clearly many of our previous problems could be solved by probing, by trial and error, and in medieval times this procedure must have been commonly used. In several problems the possibilities are rather limited so that not many attempts need to be made. We have already mentioned that a method for solving linear indeterminate problems was found quite early by the Hindu school of mathematics. In Europe a corresponding method was not discovered until a millenium later, and the date of rediscovery can be fixed quite accurately. In 1912 there appeared in Lyons a collection of ancient puzzles under the title: *Problèmes plaisans et delectables, qui se font par les nombres.* The author was Claude-Gaspar Bachet, Sieur de Méziriac (1581–1638), a gentleman, scholar, poet, and theologian, ardently devoted to classical learning. His work proved popular and a second enlarged edition appeared in 1624. Here one finds for the first time his rules for solving indeterminate problems.

One of Bachet's problems runs about as follows:

8. A party of 41 persons, men, women, and children, take part in a meal at an inn. The bill is for 40 sous and each man pays 4 sous, each woman 3, and every child $\frac{1}{3}$ sou. How many men, women, and children were there?

176

In this case we have the equation

$$x + y + z = 41, \qquad 4x + 3y + \frac{z}{3} = 40 \cdot \qquad (6\text{-}8)$$

Bachet's procedure unfortunately is complicated by a lack of algebraic symbolism. In the following sections we shall make a more systematic study of the linear indeterminate problems. Here we prefer to present a method of repeated reductions that is easy to explain. It works quite well when the numbers involved are not too large, as, for instance, in most of the examples we have already mentioned. This method was used extensively by Euler in his popular *Algebra* (1770), which devotes much space to indeterminate problems.

We shall deal first with a single linear equation

$$ax + by = c \qquad (6\text{-}9)$$

in two unknowns. As a preliminary example we take simply

$$x + 7y = 31 \qquad (6\text{-}10)$$

which may be written

$$x = 31 - 7y \qquad (6\text{-}11)$$

This shows that any integral value substituted for y in (6-11) will give an integral value for x; for instance, $y = 6$, $x = -11$; or $y = 0$, $x = 31$. Thus there will be an infinite set of pairs of solutions. But if one requires positive solutions, one must have both $y > 0$ and

$$x = 31 - 7y > 0$$

thus $y < 4\frac{3}{7}$. This gives only four possibilities, $y = 1$, 2, 3, 4, with the corresponding values, $x = 24$, 17, 10, 3, for the other unknown.

This trivial example was introduced in order to show that when one of the coefficients of x and y in (6-9) is unity, as in (6-10), the solution is immediate. The guiding principle in the method used below is to reduce the more general equations in successive steps to this simple form. The first example given by Euler is:

Write the number 25 as the sum of two (positive) integers, one divisible by 2 and the other by 3.

The two summands may be taken to be $2x$ and $3y$ so that

$$2x + 3y = 25 \qquad (6\text{-}12)$$

is the equation to be fulfilled. Since x has the smaller coefficient,

177

we solve for x and find by taking out the integral parts of the fractional coefficients

$$x = \frac{25 - 3y}{2} = 12 - 2y + \frac{1 + y}{2} \qquad (6\text{-}13)$$

Because x and y are integers, the quotient

$$t = \frac{1 + y}{2} \qquad (6\text{-}14)$$

is integral. Conversely, any integral value t we may give to this quotient (6-14) will make y integral

$$y = 2t - 1$$

and also x integral according to (6-13)

$$x = 12 - 2y + t = 14 - 3t$$

This shows that the general integral solution of (6-12) is

$$x = 14 - 3t, \qquad y = 2t - 1$$

and one can verify by substitution that they actually satisfy the equation. Consequently, there is an infinite number of solutions, one for each integral t. For instance, when $t = 10$, $x = -16$, $y = 19$.

But if one is limited to positive values, one must have

$$x = 14 - 3t > 0, \qquad y = 2t - 1 > 0$$

hence

$$\tfrac{1}{2} < t < 4\tfrac{2}{3}$$

and there are only four permissible values, $t = 1, 2, 3, 4$. The corresponding solutions are

$$x = 11, 8, 5, 2$$

$$y = \ \ 1, 3, 5, 7$$

This gives the decompositions

$$25 = 22 + 3 = 16 + 9 = 10 + 15 = 4 + 21$$

required in the original problem, as one could have verified without much effort by probing.

This example requires only one reduction. In most cases two or more steps are required. Let us illustrate this by another example taken from Euler's *Algebra*.

178

A man buys horses and cows for a total amount of $1,770. One horse costs $31 and one cow $21. How many horses and cows did he buy?

When x is the number of horses and y the number of cows, the condition

$$31x + 21y = 1{,}770 \qquad (6\text{-}15)$$

must be fulfilled. Here y has the smaller coefficient so we solve for y and find

$$y = \frac{1{,}770 - 31x}{21} = 84 - x + \frac{6 - 10x}{21} \qquad (6\text{-}16)$$

This requires that the quotient

$$t = \frac{6 - 10x}{21}$$

shall be integral. Our task is, therefore, to find integers x and t such that

$$21t + 10x = 6 \qquad (6\text{-}17)$$

As can be seen, this is an equation of the same type as (6-15) but with smaller numbers so that a first reduction has been performed. Since x has the smaller coefficient, we derive from (6-17)

$$x = -2t + \frac{6 - t}{10} \qquad (6\text{-}18)$$

We conclude that x can only be integral when

$$u = \frac{6 - t}{10}$$

is integral or

$$t = 6 - 10u$$

for some integer u. By substituting this value into (6-18) and then x into (6-16), one finds

$$x = -12 + 21u, \qquad y = 102 - 31u$$

Any integral value of u will give integers x and y satisfying the equation (6-15) so that we have obtained the general solution. The form of the problem requires, however, that x and y must be positive. This leads to the conditions

179

$$-12 + 21u > 0, \qquad 102 - 31u > 0$$

or

$$\tfrac{12}{21} < u < 3\tfrac{9}{31}$$

There are, therefore, three possible values $u = 1, 2, 3$, and the corresponding solutions are

$$x = 9, 30, 51$$
$$y = 71, 40, 9$$

We could have made the solution of the problem unique, for instance, by requiring in the formulation that the number of horses would be greater than the number of cows.

As a last example of this type we shall take problem 4 in the preceding section, stated by Bhaskara. It is of interest because it permits us to mention some simplifications that often are available in the solution of indeterminate problems.

We observe first that in (6-4) the coefficients 221, 65, and 195 are all divisible by 13. This factor can, therefore, be canceled and the equation becomes

$$17x + 5 = 15y \qquad (6\text{-}19)$$

Here, furthermore, both 5 and $15y$ are divisible by 5 so that $17x$ must have this factor. But 17 is prime to 5 so that x must be divisible by 5, and we can write

$$x = 5x_1$$

When this is substituted in (6-19), one can cancel by 5 and have the still simpler equation

$$17x_1 + 1 = 3y$$

By writing this

$$y = \frac{17x_1 + 1}{3} = 6x_1 + \frac{1 - x_1}{3}$$

we see that

$$\frac{1 - x_1}{3} = t$$

is integral. This gives $x_1 = 1 - 3t$ and

$$x = 5x_1 = 5 - 15t, \qquad y = 6 - 17t \qquad (6\text{-}20)$$

as the general solution.

Let us ask for the positive solutions. One obtains, as previously, the conditions

$$5 - 15t > 0, \qquad 6 - 17t > 0$$

or

$$t < \tfrac{1}{3}, \qquad t < \tfrac{6}{17}$$

This shows that all values $t = 0, -1, -2, \cdots$ will give positive solutions in (6-20). To obtain positive values for this parameter or auxiliary variable, it is convenient in (6-20) to write $t = -u$ so that

$$x = 5 + 15u, \qquad y = 6 + 17u$$

becomes the general solution; all values $u = 0, 1, 2, \cdots$ give positive answers, namely,

$$x = 5, 20, 35, 50, \cdots$$
$$y = 6, 23, 40, 57, \cdots$$

This example illustrates the fact that even when the solutions are required to be positive there may be an infinite number of them.

Problems.

1. Divide 100 into two summands such that one is divisible by 7, the other by 11. (Euler.)

2. Required, such values of x and y in the indeterminate equation

$$7x + 19y = 1{,}921$$

that their sum $x + y$ may be the least possible. (From Barlow, *An Elementary Investigation of the Theory of Numbers,* etc., London, 1811.)

3. In the forest 37 heaps of wood apples were seen by the travelers. Atfer 17 fruits were removed the remainder was divided evenly among 79 persons. What is the share obtained by each? (Mahaviracarya.)

4. Find two fractions having 5 and 7 for denominators whose sum is equal to $\tfrac{26}{35}$. (Barlow.)

5. A party of men and women have paid a total of 1,000 groschen. Every man has paid 19 groschen and every woman 13 groschen. What is the smallest number of persons the party could consist of? (Modified from Euler.)

181

The Game of "As If"

by Irving Adler

*Mathematical discovery is a special example of scientific discovery.
The nature of the thinking used in scientific discovery is explained in
the next selection, taken from Irving Adler's book,*
Logic for Beginners.

THE CLOSED BOX

The diagram below represents a box that is completely
closed except for two tiny holes at the bottom. Two equal weights,
A and B, are hanging from ropes threaded through these holes.
*Suppose that each of the weights A or B can be pulled down, and
that, when one weight is pulled down any distance, the other
weight moves up an equal distance.* The object of the game is to
explain why this happens, without opening the box to see what is
inside. To explain the behavior of the weights we try to imagine a
hookup of ropes and pulleys inside the box that would make the
weights behave the way they do. Then we can say the weights
behave *as if* there were such and such a hookup inside the box.

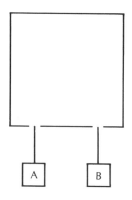

There are many possible solutions to this game of "as if." One of them is shown in the drawing below:

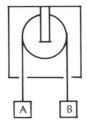

1. Suppose that, *when A is pulled down any distance, B moves up twice that distance.* Draw a hookup of ropes and pulleys that would explain this behavior of the weights.
2. Suppose that, *when B is pulled down any distance, A moves up three times that distance.* Draw a hookup of ropes and pulleys that would explain this behavior of the weights.
3. Suppose that, *when B is pulled down any distance, A moves down twice that distance.* Draw a hookup of ropes and pulleys that would explain this behavior of the weights.

(See page 187 for possible solutions.)

THE PERISCOPE

A periscope is a device by which you can see in one direction by looking in another direction. Each of the boxes below represents a different kind of periscope. The eye looks into the box through one hole, and sees things outside the box through the other hole, in the direction indicated by the arrow. In each case draw an arrangement of mirrors inside the box that would explain why the box behaves the way it does.
(See page 187 for possible solutions.)

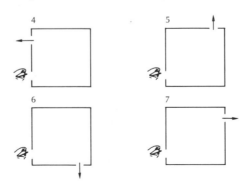

"DROODLES"

A popular game consists of showing someone a drawing made of a few simple lines and asking him what it is a picture of. The drawing below is a typical example. The expected answer is, "a soldier and a dog going around a corner." The game is really

a variety of the "as if" game, because the answer means, "it looks *as if* a soldier and a dog have just gone around a corner, and we see the bayonet of the soldier and the tail of the dog just before they disappear from view." The cartoonist Roger Price called drawings like these *droodles*. Some droodles are shown below. See if you can tell what each "droodle" is a picture of.

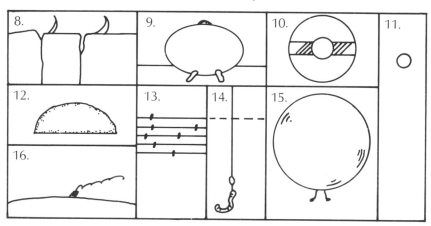

"AS IF" IN DAILY LIFE

The game of "as if" is not merely a game. It is an important method of thinking that we use all the time in our daily lives. We use it whenever we see something and try to figure out what must have happened to bring about what we see.

(See page 187 for possible solutions.)

A scene is shown in each of the drawings on page 185. Express your judgment of what caused the scene by completing the sentence "It looks as if . . ." (Write your answers on a separate sheet of paper.)

184

MAKING UP REASONS

The game of "as if" is the opposite of the game "draw your own conclusion." In the game "draw your own conclusion," as we played it on page 114*, we are given some *reasons,* and we look for a *conclusion* that follows from the reasons. In the game of "as if," we are given a *conclusion* consisting of the scene or events that we observe, and we try to make up *reasons* that will explain the *conclusion.* For example, to explain why the man shown in drawing number 17 is so wet, we may make up these reasons: It is raining outside. The man was caught in the rain, and did not have an umbrella.

Somebody else looking at drawing number 17 may make up entirely different reasons to explain why the man is wet. He may say, for example, that the man is wet because his neighbor was watering his garden, and, turning the garden hose without looking, he accidentally poured water all over him. Both sets of reasons serve to explain what we see in the drawing. Which set of reasons is correct? We cannot tell from what we see in the drawing alone. We need more information before we can reject one or the other set of reasons as being wrong.

(See page 187 for possible solutions.)

*Page 114 of Irving Adler's book, *Logic for Beginners*. 1964. John Day.

A set of reasons used to explain a given set of facts is called a *hypothesis*. When a detective tries to solve a crime, first he gathers all the facts that he can that relate to the crime. These are his "clues." Then he makes up several hypotheses, each of which could explain the facts. Then he sets out to eliminate some of the hypotheses. To eliminate some of these hypotheses he follows this procedure: First he draws as many conclusions as he can from each hypothesis. Second he looks for new facts that will either agree with or contradict these new conclusions. If the new facts contradict any conclusions drawn from a hypothesis, then the hypothesis is false, and it must be rejected. If the new facts agree with conclusions drawn from the hypothesis, then they are additional evidence in favor of that hypothesis. If the detective starts with three hypotheses and he proves two of them are false, he is still not sure that the third one is true. But at least he knows that it is the only one that *may* be true. He hopes that, as he continues his investigation, he will turn up some facts that will prove beyond doubt that the third hypothesis really is true.

HYPOTHESES IN SCIENCE

The work of a scientist is like the work of a detective trying to solve a crime. The result of an experiment is the "crime" that the scientist is trying to solve. The measurements he makes in the experiment are his "clues." The scientist makes up a hypothesis, which is a set of reasons that can explain what happened in the experiment. The scientist next draws additional conclusions from the hypothesis. These new conclusions are called "predictions." Then he performs new experiments to see if the predictions are true. If the predictions made from a hypothesis turn out to be false, then the hypothesis is false. If the predictions turn out to be true, they are additional evidence in favor of the hypothesis. If all the predictions made from a hypothesis turn out to be true, then the hypothesis is called a "scientific law."

TWO KINDS OF REASONING

The kind of reasoning that is used in the game of "draw your own conclusion" is called *deductive reasoning*. The kind of reasoning that is used in the game of "as if" is called *inductive reasoning*. Deductive reasoning and inductive reasoning are the chief methods we use for figuring out new knowledge from old knowledge.

186

Solutions to Puzzles

(Note: One solution is given for each of the puzzles 1 through 20. Other solutions are also possible.)

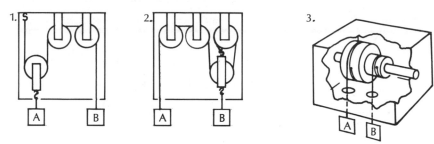

In answers 4 to 7 the arrow shows the direction in which the eye sees. M indicates a mirror.

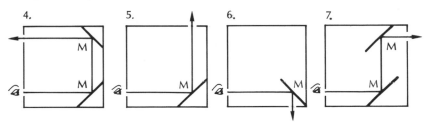

8. Two cats going over a fence. 9. A woman on her knees washing the floor. 10. The old oaken bucket seen from the bottom of the well. 11. A motorcycle coming toward you on a dark road at night. 12. A wildcat in a cave. 13. Birds sitting on telephone wires. 14. A worm on a fish-hook in water. 15. A little boy blowing up a big balloon. 16. A steamship below the horizon. 17. It looks as if the man was caught in the rain. 18. It looks as if the cat knocked the vase off the book case. 19. It looks as if the man was tripped by the dog chasing the cat. 20. It looks as if the cat knocked the milk bottle over and drank some of the spilled milk.

To Pergamon Press Ltd, England for "Fibonacci Numbers" from the English language edition Vorob'ev "Fibonacci numbers," translated by Halina Moss and edited by Sneddon.

To Marie Rodell for the selection from *The Sea Around Us* by Rachel L. Carson. Copyright ©1950, 1951, 1961 by Rachel L. Carson. Reprinted by permission of the Literary Agent.

To United Press International for the article on the 1970 draft lottery as it appeared in the Bennington (Vermont) Banner of July 1, 1970.

To University of California Press for selections from *Mathematical Principles of Natural Philosophy and His Systems of the World* by Sir Isaac Newton, tr. revised by Florian Cajori. Reprinted by permission of the Regents of the University of California.

To Holt, Rinehart and Winston, Inc. for "The Sea Monster." From *Twenty Thousand Leagues Under the Sea* by Jules Verne. Copyright 1932, ©1960 by Holt, Rinehart and Winston, Inc. Reprinted by permission of Holt, Rinehart and Winston, Inc.

Scholastic Book Services for the chapter "Isaac Newton 'All Was Light'." Reprinted by permission from *Breakthroughs in Science* by Isaac Asimov, ©1959 by Scholastic Magazines, Inc.

To School Science and Mathematics Association, Incorporated for material from the article "Hamming It Up Mathematically" by Dale M. Shafer. Reprinted by Permission of the Editor of *School Science and Mathematics*, the Journal of the School Science and Mathematics Association.

Photographs

Brown Brothers   151
Planning Research Corporation   52
Planning Research Corporation   164
United Press International   145

BCDEFGHIJ 07987654

Printed in the United States of America